Sustainable Wastewater Management in Developing Countries

Other Titles of Interest

Advances in Water and Wastewater Treatment, **edited by Rao K. Surampalli and K. D. Tyagi** (ASCE Committee Report, 2004). Presents state-of-the-art information on the application of innovative technologies for water and wastewater treatment with an emphasis on the scientific principles for pollutant or pathogen removal. (ISBN 978-0-7844-0741-7)

Field Guide to Environmental Engineering for Development Workers: Water, Wastewater, and Indoor Air, **by James R. Mihelcic, Lauren M. Frye, Elizabeth A. Myre, Linda D. Phillips, and Brian D. Barkdoll** (ASCE Press, 2009). Explains sustainable engineering techniques for application in preparing for and executing international engineering service projects. (ISBN 978-0-7844-0985-5)

GIS Tools for Water, Wastewater, and Stormwater Systems, **by Uzair M. Shamsi** (ASCE Press, 2002). Offers guidelines to develop GIS applications for water, wastewater, and stormwater systems. (ISBN 978-0-7844-0573-4)

Treatment System Hydraulics, **by John Bergendahl** (ASCE Press, 2008). Addresses the nuts-and-bolts of typical treatment systems, examines typical variables, and describes methods for solving problems encountered in the field. (ISBN 978-0-7844-0919-0)

Water Resources Engineering: Handbook of Essential Methods and Design, **by Anand Prakash** (ASCE Press, 2004). Proposes practical methods to solve problems commonly encountered by practicing water resources engineers in day-to-day work. (ISBN 978-0-7844-0674-8)

Sustainable Wastewater Management in Developing Countries

New Paradigms and Case Studies from the Field

Carsten Hollænder Laugesen and Ole Fryd

with Thammarat Koottatep and Hans Brix

Library of Congress Cataloging-in-Publication Data

Laugesen, Carsten Hollander.
 Sustainable wastewater management in developing countries : new paradigms and case studies from the field / Carsten Hollander Laugesen and Ole Fryd ; with Thammarat Koottatep and Hans Brix.
       p. cm.
 Includes bibliographical references and index.
 ISBN 978-0-7844-0999-2
 1. Sewage disposal—Developing countries. 2. Sewage—Purification—Developing countries. 3. Water reuse—Developing countries.   I. Fryd, Ole.   II. Title.

TD627.L38 2009
628.3'62091724—dc22
                          2009031915

Published by American Society of Civil Engineers
1801 Alexander Bell Drive
Reston, Virginia 20191
www.pubs.asce.org

Any statements expressed in these materials are those of the individual authors and do not necessarily represent the views of ASCE, which takes no responsibility for any statement made herein. No reference made in this publication to any specific method, product, process, or service constitutes or implies an endorsement, recommendation, or warranty thereof by ASCE. The materials are for general information only and do not represent a standard of ASCE, nor are they intended as a reference in purchase specifications, contracts, regulations, statutes, or any other legal document.

ASCE makes no representation or warranty of any kind, whether express or implied, concerning the accuracy, completeness, suitability, or utility of any information, apparatus, product, or process discussed in this publication, and assumes no liability therefor. This information should not be used without first securing competent advice with respect to its suitability for any general or specific application. Anyone utilizing this information assumes all liability arising from such use, including but not limited to infringement of any patent or patents.

ASCE and American Society of Civil Engineers—Registered in U.S. Patent and Trademark Office.

Photocopies and reprints. You can obtain instant permission to photocopy ASCE publications by using ASCE's online permission service (http://pubs.asce.org/permissions/requests/). Requests for 100 copies or more should be submitted to the Reprints Department, Publications Division, ASCE (address above); e-mail: permissions@asce.org. A reprint order form can be found at http://pubs.asce.org/support/reprints/.

Copyright © 2010 by the American Society of Civil Engineers.
All Rights Reserved.
ISBN 978-0-7844-0999-2
Manufactured in the United States of America.

17 16 15 14 13 12 11 10   1 2 3 4 5

Table of Contents

Preface vii

1 **Sustainable Wastewater Management:
 An Introductory Overview** 1
  1.1 At a Crossroad 1
  1.2 An Issue of Global Importance 7
  1.3 The Way Forward 9

2 **Reflections on Sustainable Wastewater Management** 19
  2.1 Discussing Appropriateness and Sustainability 19
  2.2 The Eleven Fundamental Issues 20

3 **Elements of Sustainable Wastewater Management** 33
  3.1 Framing Appropriateness and Sustainability 33
  3.2 The Ten Nods of Appreciation 34
  3.3 Scale, Systems, and the Six Elements for Appropriateness ... 39
  3.4 Element 1: Wastewater Collection Systems 41
  3.5 Element 2: Wastewater Treatment Systems 45
  3.6 Element 3: Energy Consumption 58
  3.7 Element 4: Urban Integration 60
  3.8 Element 5: Re-Use and Re-Entry of Wastewater 62
  3.9 Element 6: Organization and Finance 70
  3.10 Nodding in Perspective: The Width and Depth of
       Assessing Appropriateness and Sustainability 80
  3.11 Sense and Simplicity 87

4 **Sustainable Wastewater Management at the Chairman's
 House: A Recovery-Based, Closed-Loop Household System** 89
  4.1 The Living Lab of Dr. Ksemsan Suwarnarat 89
  4.2 Reflections on Appropriateness and Sustainability 92
  4.3 Smart Technologies at the Chairman's House 96

5 **Constructed-Wetland Wastewater Treatment at Baan
 Pru Teau: A Low-Cost Cluster Community System** 100
  5.1 Supporting a Cluster of Houses 100
  5.2 Reflections on Appropriateness and Sustainability 104
  5.3 Smart Technologies at Baan Pru Teau 110

6	**Wastewater Management Design at Koh Phi Phi: A Recovery-Based, Closed-Loop System**	**114**
	6.1 The Flower and the Butterfly	114
	6.2 Reflections on Appropriateness and Sustainability	124
	6.3 Smart Technologies on Koh Phi Phi	132
7	**Energy-Optimized Wastewater Treatment at Siriraj Hospital: A Large-Scale, On-Site Treatment System**	**152**
	7.1 Year after Year	152
	7.2 Reflections on Appropriateness and Sustainability	156
	7.3 Smart Technologies at Siriraj Hospital	159
8	**Constructed Wetland at Patong: A River Treatment System**	**163**
	8.1 Doing the Next Best	163
	8.2 Reflections on Appropriateness and Sustainability	169
	8.3 Smart Technologies in Patong	173
9	**Pond and Constructed-Wetland Treatment at Sakon Nakhon: A Sustainable Municipal System**	**178**
	9.1 Fields of Action	178
	9.2 Reflections on Appropriateness and Sustainability	185
	9.3 Smart Technologies at Sakon Nakhon	188
10	**Wastewater Planning in Pathumthani Province: Appropriate Planning of Large-Scale Wastewater Management**	**194**
	10.1 Thinking Small, Big Scale	194
	10.2 Reflections on Appropriateness and Sustainability	223
11	**The Way Forward, Mainstreaming, and Other Reflections**	**228**
	11.1 The Sustainable Ecosystem Approach: Going Mainstream	229
	11.2 Three Key Interlinked Conclusions Are Mainstreamed	231
	11.3 Local Context: Going Mainstream	**235**

References	239
Index	241
About the Authors	251

Preface

When did I realize change was needed within the wastewater management sector in the tropics? Some years ago I worked in Malaysia as an advisor to a state environmental protection department. We wanted to draft a regulation for wastewater treatment and therefore decided to evaluate the status of the existing systems. We found that 91 treatment plants had been constructed in the state. Of these, 89 were either malfunctioning or inactive. Of the two in operation, one was a recently built plant that was expected to fail soon. The other was a municipal pond system that was not operated by anyone but, because wastewater was flowing through it by gravity, it was therefore labeled as being in operation!

Some years later I moved on to work in Thailand in the newly established national wastewater management authority. One of the first things I did was to visit most of the 76 municipal treatment plants in the country, and it quickly became apparent that the situation in Thailand was similar to that in Malaysia. All these expensive engineered treatment systems, and almost all of them malfunctioning! It really puzzled me. As a public sector specialist I had never seen anything like this before—a sector where, apparently, the same mistakes were repeated over and over again. How could a sector and its planners, engineers, and economists accept, or at least not constantly challenge, such a degree of failure? How could they continue to propose, design, and finance similar systems and technologies that already had been proven not to work? Despite how amazing this looked, I came to realize it was nevertheless a fact. I consulted other colleagues within the sector who had worked outside the luxurious conditions of Western countries, and they all had the same stories and experiences—in Asia, Africa, and South America, in fact, in all tropical countries. Because I have now worked for more than 10 years in developing countries, mainly in Asia and South Africa, I, too, have come to know all the complex and interlinked political, economic, cultural, and institutional reasons why wastewater management in developing countries lacks the ability to change direction and approaches.

—*Carsten H. Laugesen*

This book presents reflections on and actual stories about appropriate and sustainable wastewater management systems in the tropics. General reflections are followed by case stories and the implications and applications that can be drawn from these stories from the field. This book is intended to inspire rather than prescribe and dictate; to support thoughtful innovations rather than replication of dogmas.

Our aim is to reflect on, discuss, and provide examples (and thereby hopefully inspire) a broader use of robust, reliable, cost-effective, and efficient wastewater management systems that work in practice. "Sense and Simplicity" is the principle we have chosen to guide theory and practice.

This book has been written by a multidisciplinary team of people who would like to support better wastewater management planning and implementation in the future. This team has experienced, especially in developing countries, numerous failures of traditional planning, design, and implementation of wastewater management systems, and would like to contribute to reducing such failures in the future. We are not locked into a single approach (e.g., "small is beautiful," "pro-low-tech," or "anti-centralization"). We believe in localized best solutions—a "fit the local context" approach to assessing what is best. Success is only achieved when something works in real life—not in theory or not what might be possible if this or that were in place. Success is what proves to work, year after year, and thus has an actual positive impact on public health and the environment.

The main authors are Carsten H. Laugesen and Ole Fryd, with Thammarat Koottatep and Hans Brix providing invaluable inputs, comments, and corrections. The following have also provided valuable contributions and comments to this book and the experiences it is based upon: Ksemsan Suwarnarat, Sarawut Srisakuna, Suchai Janepojanat, Chatdanai Jiradecha, Niras Limprayoonyong, Pisit Srivilairit, Henrik Lynghus, Jacob Hamburger Hansen, Ejlif Mikkelsen, Kenneth Wright, Mikkel Rye Christensen, Bablu Virinder Singh, Tony Greer, Waraporn Kanchanapiboon, Thasanee Dejpraikhala, and Kitti Uyakul. We are deeply grateful to the Danish International Development Agency (DANIDA), especially Kit Clausen and Marinette Forbes Ricarde, and to Tracy Hart from the World Bank, for their valuable support.

—*Carsten H. Laugesen, Ole Fryd,*
Thammarat Koottatep, and Hans Brix

1

Sustainable Wastewater Management: An Introductory Overview

))) 1.1 At a Crossroad

Wastewater management in developing countries is at a crossroad, and it is generally agreed there is an urgent need for a shift in the approach to wastewater management and planning in developing countries. Needs are growing, resources are scarce, previous management systems have failed, and traditional techniques and solutions are not rapid, efficient, or cost-effective enough to solve the wastewater management problems developing countries are facing.

At a time when traditional paradigms have proven insufficient and new ones have yet to fully take form, many new emerging views, opinions, and competing systems and technologies are seeing the light of day. Some of these are more appropriate and sustainable than others. This is an excellent time for rethinking, experimenting, and seeking new paths.

1.1.1 It Is Difficult to Change Direction

Despite the past failure of most centralized systems, it is likely that most new wastewater management systems in developing countries will continue to be advanced, centralized, and with a continued high probability for failure. The reasons for this are many and interlinked.

The first and probably most important reason is the political preference for large, one-off investments. Other significant reasons include inertia; the wish to compare favorably with developed countries; the education and expertise of local wastewater engineers; and whether international water and wastewater consortia are providing funding and consultancy.

The complexity of wastewater planning often supports the choice of advanced, centralized wastewater management systems. When planning large-

> **Box 1-1. Wastewater Management at a Crossroad**
>
> Thailand's 1998 national wastewater treatment plan recommended that more than two-thirds of the country's wastewater be treated in centralized activated sludge treatment plants. The largest system was designed to treat more than half a million cubic meters of wastewater per day. Based on this plan, in the last decade Thailand implemented 76 centralized wastewater management systems at a cost of approximately $2 billion USD. These centralized treatment plants—activated sludge plants, stabilization ponds, and aerated lagoons—were intended to treat 20% of the wastewater produced in the country (Fig. 1-1).
>
> However, those 76 wastewater treatment systems have had a discouragingly low impact; the effects of the activated sludge treatment plants have actually been disastrous. Very few—perhaps less than five treatment facilities—are effectively in operation today. As a result, Thailand has tabled all new investments in wastewater management systems and now must decide whether it will continue the implementation of capital-intensive, centralized advanced wastewater management systems, or whether other methods are more feasible. Issues of sustainability of technology, urgency, time span for implementation, costs, financing, fee structure, water quality standards, organizational setup, and national policies are all up for discussion and are more or less undecided in Thailand, as in many other developing countries.

scale wastewater management systems, the perception is often that it is more complex to design, implement, and manage *decentralized* wastewater management systems for large areas or quantities of wastewater. For conventional centralized systems, planners can often just refer to a textbook. Even though decentralized on-site treatment and cluster management systems (e.g., constructed wetlands) can treat large amounts of wastewater at low cost, they require careful and tedious planning with respect to local conditions in each individual case. Decentralized systems cannot be constructed based on standard textbooks, as will be illustrated in more detail in Chapter 11.

Nearly all wastewater in Europe and North America is managed by full-scale trunk sewers and large, centralized wastewater treatment facilities. In all other regions of the world, centralized sewer systems account for less than half of all wastewater management systems (Fig. 1-2). This is normally interpreted as a gap between developed and developing countries, with the obvious but faulty conclusion that developing countries need to implement more centralized waste management systems.

Figure 1-1. Wastewater treatment plants in Thailand. *Clockwise from top left:* a pond system in Sakon Nakhon; a man fishing in a stalled treatment unit in Samui; a denitrification system in Patong; and an aeration tank in operation in Prachinburi.

Not just one solution exists for all technical problems. The application of huge, centralized systems in developing countries—collecting both wastewater and surface water for treatment at large and advanced treatment facilities—has largely precluded the testing and application of alternative, decentralized wastewater management systems in developed countries. Most developed countries today are bound to centralized systems because developing alternate systems is prohibitively expensive. Likewise, these centralized systems are subject to enormous replacement costs at frequent intervals. Typically, sewers, manholes, and other technical facilities must be replaced every 50 to 100 years, and this day is approaching in many developed countries. This is causing growing concern over the budgets required for these huge replacement tasks.

Conversely, most developing countries have not yet made investments in centralized sewer and treatment systems, giving these regions opportunities for experimenting with new and perhaps more suitable short- and long-term concepts for wastewater management. The percentage of urban houses

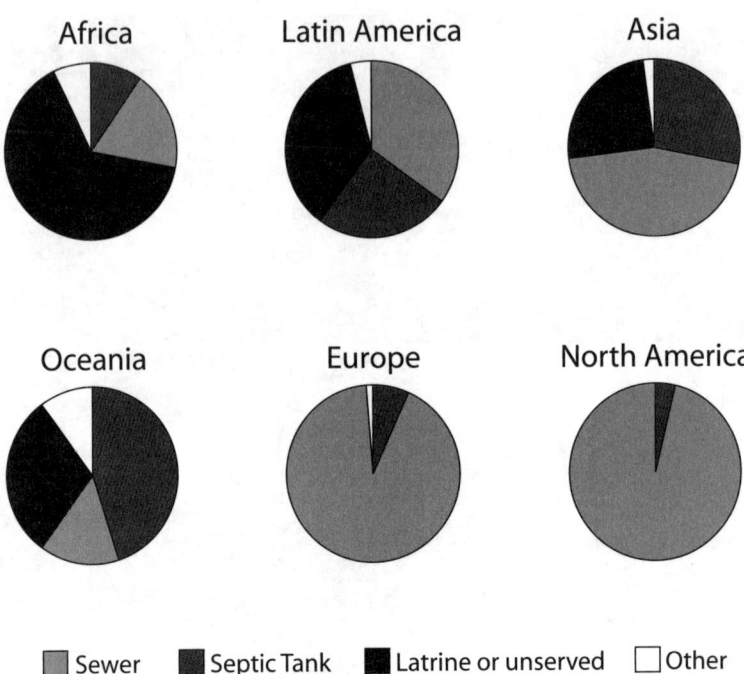

Figure 1-2. Type of sanitation systems by world region.
Source: Adapted from WHO and UNICEF (2000).

served by on-site sanitation installations in developing cities and countries is, as Fig. 1-3 shows, still high (50% to 90% of the urban populations). Alternative solutions or, preferably and more precisely, the mixing and matching of different wastewater management solutions are still possible in most developing countries and cities.

Today, many areas in developing countries would not be able to provide the water supply service level required for waterborne sanitation. The lowest coverage of drinking water services is found in the low-income developing countries as well as in the poorer areas in middle-income countries. Ironically, if these countries succeed in providing their entire population with safe, centralized drinking water services and implementing centralized collection systems, the human waste (which previously was contained and treated via on-site technology) would appear as wastewater pollution in nearby coastal waters, threatening the coastal ecosystems. From a public health point of view, a valid question is whether it is wise to greatly dilute pathogens, which originally are produced in compact and manageable form.

Malfunctioning wastewater management systems in developing countries have become a growing concern during recent decades. Historical, politi-

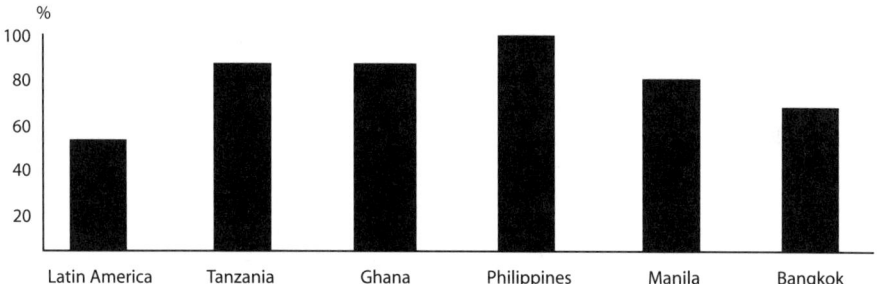

Figure 1-3. Proportion of urban populations served by on-site sanitation systems.
Source: Adapted from Strauss et al. (2000).

cal, economic, organizational, traditional, and cultural reasons have all linked to create a wastewater sector under pressure; we discuss in detail some of the key reasons for this in Chapter 2.

Professionals working with wastewater management systems in developing countries are at a crossroad. Most realize something must change, but change to what is still not fully obvious.

> ###))) Box 1-2. A Shift in Professional View?
>
> Sometimes we must change our views and practice. If something doesn't work, it doesn't work, regardless of how many excuses we make or how many times we say, "If this or that were in place, it would work." When and why do we change our professional minds? Thomas Kuhn, in his famous book *The Structure of Scientific Revolutions* (Kuhn 1962), used the term "paradigm shift" for when previous understandings are changed to new understandings and ways of dealing with problems and solutions. A paradigm shift describes a change in basic assumptions within a profession. Kuhn uses the duck/rabbit optical illusion (Fig. 1-4) to demonstrate how a paradigm shift can cause one to see the same information in an entirely different way.
>
> According to Kuhn, a paradigm shift occurs when practitioners encounter anomalies that cannot be explained by the universally accepted paradigm. Some anomalies are brushed away as acceptable levels of error, or are simply ignored. To put this in the context of wastewater management in developing countries, some practitioners accept numerous anomalies, or malfunctions, and still argue for the continued application of the current paradigm. But, according to Kuhn, anomalies have various levels of significance to the practitioners at a given time; when enough significant anomalies have accrued

against a current paradigm, the discipline is thrown into a state of crisis. During this crisis, new ideas—perhaps ones previously discarded—are tried. Eventually a new paradigm is formed that gains its own new followers, and an intellectual battle takes place between the followers of the new paradigm and the hold-outs of the old paradigm. Kuhn pointed out that sometimes the convincing force is just time itself and the human toll it takes. He quotes Max Planck: "A new scientific truth does not triumph by convincing its opponents and making them see the light, but rather because its opponents eventually die, and a new generation grows up that is familiar with it" (Kuhn 1962).

How are professionals' views changed? A professional embraces a new view for all sorts of reasons, including rational considerations, personality, nationality, and the reputations of prior innovators. But if a professional is to decide to change a way of practice, that person must have faith that the new paradigm will succeed with the many large problems that confront it. This faith is based on future promise rather than on past achievement. *Faith* is therefore the operative word—we still do not know whether the new approaches will work. Sometimes this faith is based on personal and inarticulate aesthetic considerations. This is not to suggest that new paradigms triumph ultimately through some mystical aesthetic. The new paradigm appeals to the individual's sense of the appropriate or the aesthetic—the new paradigm is said to be neater, more suitable, more sustainable, simpler, or more elegant.

Professionals solve problems concerning the behavior of nature and society. Although the concerns may be global, the problems are matters of detail, and the solutions that satisfy a professional must satisfy the society and the community.

Figure 1-4. The duck/rabbit illusion.

Source: Jastrow (1899).

⟩⟩⟩ 1.2 An Issue of Global Importance

In our opinion, what is required today is a new way at looking at problems and solutions within the wastewater management sector in developing countries. The growing number of malfunctioning centralized or advanced wastewater management systems in developing countries, and the lack of agreed-upon alternatives, is unfortunately not just a professional, technical problem for wastewater engineers. It is an issue of enormous global importance: 2.6 billion people—more than 40% of the world's population—are today living without adequate management of the wastewater they produce (WHO/UNICEF 2006).

The United Nations Millennium Development Goals call for halving, by 2015, the proportion of people without sustainable access to safe drinking water and basic sanitation. To meet this target, more than 1.6 billion additional people need to gain access to improved sanitation over the coming decade; this will require more than 100,000 new house installations every day until 2015 (WHO/UNICEF 2006). In June 2006, the Copenhagen Consensus Center ranked community-managed water supply and sanitation *second* among the 40 most important challenges for the global community, after improved basic health services to fight communicable diseases (CCC 2006).

The importance becomes clear when the impact of inappropriate wastewater on public health is studied. To put it bluntly, lack of appropriate and sustainable wastewater management systems kills people. Each year, more than 2.2 million people die from water- and sanitation-related diseases (WHO/UNICEF 2000). In developed countries, by far the main cause of death is circulatory diseases (75%), resulting from too much and unhealthy food combined with too little exercise. In developing countries, the main causes of death are primarily infectious and parasitic diseases (43%); poor management of nearby environments; and food, water, wastewater problems combined with inadequate public health services. In developing countries this only accounts for 1.2% of deaths (Fig. 1-5). Infectious and parasitic diseases linked to contaminated water is the third leading cause of productive years lost to morbidity and mortality in the developing world (WHO 2003). Diarrheal death rates are typically about 60% lower among children living in households with adequate water and sanitation facilities than among those in households without such facilities.

A WHO survey in 63 developed and developing countries distinguished between the type of sanitation services reaching the upper- and lower-income urban populations, and showed huge differences in the provision of sanitation and wastewater services between urban low- and high-income areas, irrespective of country location. Sustainable wastewater management is largely a poverty issue (WHO/UNICEF 2000).

Figure 1-5. Causes of death: developed and developing world.
Source: Adapted from WHO (1997).

Clean drinking water and good wastewater management are high on the priority lists of local municipalities and communities, as surveys and actual local resource prioritizations have repeatedly shown. Budgets for infrastructure and environment are used for (in descending order) providing clean drinking water; providing electricity; properly diverting stormwater; getting rid of wastewater; and getting rid of rubbish. Clean rivers and seas, integrated river basin management, environmental impact assessments, environmental indicators—all these environmental management newcomers enter the priority lists when, and only when, the issues of water, electricity, and waste already have been dealt with.

Approximately half the world's population has no hygienic means of disposing of sanitary wastewater from toilets, and an even greater number lack adequate means of disposing of wastewater from kitchens and baths ("greywater"). Wastewater management is important primarily because it saves lives, but at certain locations at certain times it is also important for other reasons. Local economy is one of them.

Wastewater management systems in developing countries are often implemented where local income is under pressure as a result of pollution from wastewater discharge. Consider, as an example, the wastewater infrastructure investments made so far in Thailand. Patong, Hua Hin, Pattaya, Koh Phi Phi, Koh Samui, and Koh Phangan—all renowned and important tourist locations—were some of the first places to have wastewater management facilities installed. Such decisions are logical and relevant because, for wastewater management systems to function in developing countries, they must be linked to perceived and visible local need. Besides protection of public health, decreasing income from, for example, tourism due to lack of proper management of wastewater is a parameter that clearly creates an incentive for improved and sustainable wastewater management systems.

Besides the direct human impacts on health and the local economy, lack of or inefficient wastewater management facilities result in polluted lakes, streams, rivers, beaches, and coral reefs. The environmental impact from wastewater discharge can be serious, especially near densely populated areas or where wastewater is discharged to minor, closed, or sensitive river or coastal ecosystems.

These environmental impacts have been the overriding focus for wastewater management in *developed* countries during recent decades, whereas the human impacts on health and income have gotten more attention in most *developing* countries. This is an important distinction because it has often been seen that when wastewater management systems have been introduced in developing countries with the sole aim to protect the environment, the systems have failed.

))) 1.3 The Way Forward

1.3.1 So Where Does This Leave Us? The Story Line of the Book

Today's urgent need for new approaches to wastewater management stems from both the number of implemented but nonfunctioning systems ("anomalies," Kuhn would have called them), and the urgency and size of the task. It makes good sense to investigate alternative concepts for wastewater management. Specifically, wastewater *management*—the sustainable management of wastewater from source to re-entry—should be the starting point for these considerations and discussions, not just the treatment aspect of the wastewater system, which has normally had all or most of the attention. Nor should the starting point be the old and nonproductive discussions on high-tech versus low-tech, or centralized versus decentralized, systems. We need to reach more populated areas, more quickly, at a lower cost, and with a higher degree of sustainability. We need a period of substantial, innovative rethinking.

The political, educational, legislative, institutional, and financial systems determine the successful implementation of centralized advanced wastewater management systems in any country. In developing countries, many of these systems are inadequate for the introduction of advanced wastewater management systems. In developed countries, these systems have been developed over many decades and even centuries, continuously becoming more complex and coherent as the public sector as a whole became economically and organizationally stronger and more transparent. Conversely, most communities in developing countries are probably not geared to operate and maintain such centralized systems. However, this might not be as much a question of preparedness for an advanced technology as a question of the *appropriateness* of such technology for developing countries. For that matter, the same question might apply to twenty-first-century developed countries with their greater emphasis on sustainable ecosystems, decentralization, and smart technologies.

In this book we will provide an overview of options for wastewater management in developing countries, to increase the understanding of how to develop more sustainable wastewater management systems; to generate greater awareness and understanding of an ecosystem approach that links wastewater management, ecosystems, health, and nutrition; to foster multi-disciplinarity in approaches to wastewater management; and to provide an overview of approaches and technologies that link ecosystem approaches to sustainable wastewater management.

1.3.2 System Thinking

This book offers an approach to wastewater management that reflects many of the changes in the field over the last decade. In the past, wastewater management focused mainly on specific public health effects, but there is now increased consideration of a wider range of effects on people and ecosystems. This book reflects a sustainable development framework which links urban and rural communities, and environmental, social, and economic concerns.

The shift in focus from individual to interconnected effects means that *system thinking* has helped to shape our approach here. This includes looking at human and natural systems and processes, and at how wastewater management fits in with and affects those systems. It is now not a matter of discarding untreated or treated wastewater into an environment that is somehow separate from the populated community. The issue is more one of designing a wastewater management system that works within the local ecosystems supporting the clean water, swimming areas, estuaries and coral reefs, and soils that everyone uses and enjoys.

Rather than overloading natural processes that purify water and maintain soils, wastewater management systems should be designed to work with rather than against these natural ecosystem processes. Understanding these processes before launching into the design of technical systems is fundamental for choosing a sustainable wastewater management system.

1.3.3 Ecosystems and Ecosystem Services

Understanding ecosystems and the services they provide to communities is essential. Different ecosystems are affected differently by discharge of wastewater, and the various ecosystems provide various services to the communities. The lack of well-managed or protected ecosystems can mean the loss of clean water and the loss of a river or marine farming industry, loss of recreational waters, or the decline of tourism. Some of the key impacts of wastewater discharge on ecosystems relate to eutrophication (the physical, chemical, and biological changes associated with enrichment of a body of water due to increases in nutrients and sedimentation, including toxic algal blooms and oxygen depletion), and health hazards due to pathogenic microorganisms. Linked to this is a greater scientific understanding of the whole nature of wastewater and its effect on ecosystems and their services (Fig. 1-6). It is not just a matter of managing the discharge of wastewater. The impact of organic material must also be managed. The natural purification processes and biogeochemical cycles provide a basis for determining what is environmentally sustainable management practice for wastewater. Discharge of wastewater into an environment exceeding the natural purification capacity of that environment results in the accumulation of organic materials (carbon), nitrogen, phosphorus, or other pollutants that cannot be absorbed by the ecosystem (the receiving environment). Accumulation of organic materials will result in a high oxygen demand that cannot be met by oxygen transfer from the atmosphere, resulting in undesirable anaerobic conditions. Accumulation of nitrogen and phosphorus from wastewater discharged into an ecosystem will result in eutrophication of estuaries and other river and coastal ecosystems.

This requires focus on not exceeding the capacity of the environment to assimilate the wastewater. Applying general standards for discharge of waste-

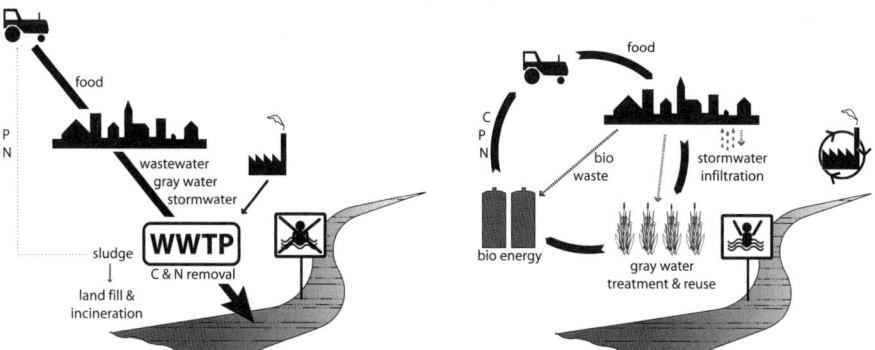

Figure 1-6. Wastewater management practices and local biogeochemical cycles. Unsustainable wastewater management practice (*shown on the left*) is not closing the local biogeochemical cycles; this results in the natural purification capacity of the receiving environment being exceeded. Sustainable wastewater management practice (*shown on the right*) is closing the local biogeochemical cycles.

Source: Adapted from Lange and Otterpohl (1997).

water in this respect is not the most appropriate way to go because each local environment has its own assimilation capacity, depending on the natural through-flow of water and climatic, vegetation, and soil conditions.

Perhaps one of the most profound changes in recent decades has been increased exploration and scrutiny of land-based wastewater treatment and re-entry systems, and a greater willingness to take creative and innovative approaches. The conventional wastewater management solution in developed countries and in rich areas in developing countries is based on the flush toilet (the flush-and-discharge model) that has been successful in disposing of wastewater for the relatively few people who have access to a regularly functioning flush toilet. This water-based model was designed and built on the premise that human waste is a waste suitable only for disposal, and that coastal environments are capable of assimilating this waste. The conventional solution for poor people in developing countries is the pit latrine (the drop-and-store model), which also has its shortcomings, especially in densely populated areas, in areas with impenetrable ground and/or high water tables, or where flooding is a problem.

These conventional, *linear* disposal solutions have led to other problems. When human waste is disposed of, nutrients and organic matter are wasted. Today there is a linear and massive flow of nutrients in the form of agricultural products from rural to urban areas, and a massive flow of nutrients, in the form of human waste and other organic matter, to rivers and coastal waters. Because human waste is regarded as a waste, its nutrients are not recycled or dedicated to productive uses on land. The linear solutions have solved some problems but also contributed to many other problems faced today: pollution of ecosystems, scarcity of water, destruction and loss of soil fertility, and lack of food security.

Until the beginning of the last century, the re-use of human waste as a fertilizer was the norm in most cultures and societies, and was an established practice in, for example, Europe and North America. Today, the challenge for sustainable wastewater management and protection of ecosystems is to regain acceptance and application of *circular* solutions for wastewater management.

1.3.4 Appropriateness and Sustainability

The emergent trends in low-cost, decentralized, nature-based infrastructure and urban wastewater management that promote the recovery and re-use of wastewater resources are extremely interesting and relevant. The concept of managing urban wastewater flows at a decentralized or intermediate level, based on microwatersheds, is similarly relevant. The concepts of use and re-use, closed-loop systems, recovery, and low energy consumption are also future-oriented, as are concepts of integrating urban planning and wastewa-

> **Box 1-3. Is a Terminology Shift Needed?**
>
> Normally, with a shift in professional attention and approaches, a change in the language and terms we use also evolves. Attention within the wastewater sector has until now almost solely been on treatment and treatment processes, and not, for example, on re-use or water utilization. This focus on treatment processes has led to a terminology primarily based on the incoming raw material and the processes, and not on the product or outcome. We use the term *wastewater treatment plants*—a term that defines what is going on inside the plant, not what comes out of the plant. One would probably never see a private company brand itself on its incoming raw material instead of its outgoing product. A furniture company is not called a wood manipulation factory; a bakery is not a flour treatment shop. But a wastewater treatment plant is called a wastewater treatment plant!
>
> Thus, there is a need for a new terminology—one that is more positive, more focused on output, and more challenging for the sector. Why do we call it "black wastewater," not "biowater" or "enriched water"? Why "septic tank," not "bioblocker" or "biocollector"? Why "sewer," not "biopipe" or "swale"? Why "sludge," not "fertilizer," "biosolid," or "biobooster"? Why "wastewater treatment plant," not "water reclamation center" or "water remediation park"? Because the power of words often determines our approach to problems and solutions, terminology always will be an important issue.

ter management strategies to conserve valuable urban resources or improve urban landscapes.

Future wastewater management systems should be recovery-based, closed-loop systems rather than traditional disposal-based, linear systems, in order to promote conservation of water and nutrient resources; to improve urban environments; and to contribute immediately and directly to public health, the local economy, and the protection of important coastal ecosystems. Systems should be fitted to the local physical, social, and institutional context and include a cultural appropriateness reflecting the local perception of, for example, soil, land, recycling, and human waste.

1.3.5 Local Context and Six Elements for Appropriateness and Sustainability

Because appropriateness and sustainability can only be understood in relation to a given setting, context, or location, the focus for wastewater management has shifted from general approaches and technologies to specific wastewater management systems for protection of ecosystems that fit into a given contextual and cultural setting at a given historical time. Local context and culture are always the base.

Besides the contextual fitness, we determine appropriateness and sustainability by assessing six elements (Fig. 1-7) that are all relevant in every wastewater management system:

1. Wastewater collection
2. Wastewater treatment
3. Urban integration
4. Energy savings
5. Re-use and re-entry
6. Organization and finance

An appropriate and sustainable wastewater management system therefore includes:

- The establishment of an *efficient wastewater collection system*
- The implementation of a *sustainable wastewater treatment facility*
- The *integration into the physical urban layout*
- *Reduced energy consumption*
- The integrated management of treated wastewater for *re-use purposes* and a *sustainable re-entry to natural waterways*
- The establishment of *sustainable organizational and financial structures* targeted to the specific task.

In Chapter 3 we will elaborate our reflections on context and these six elements of sustainability.

1.3.6 Ten Guiding Principles

We have defined 10 principles for appropriateness and sustainability of wastewater management systems in developing countries. They are:

1. Collection and treatment is undertaken on-site.
2. The collection system is short, closed, and separated from other sources.

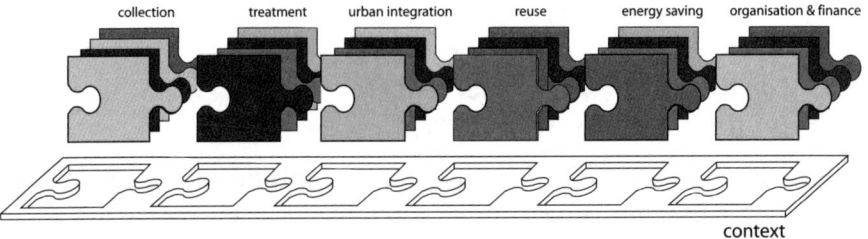

Figure 1-7. Overview of the six elements of appropriateness.

3. The effectiveness of the collection system is optimized.
4. The treatment system is the most appropriate for this location, this type of wastewater, and the resources available.
5. Smart technologies are utilized.
6. The treated wastewater is re-used.
7. Energy consumption is kept to a minimum.
8. The collection, treatment, or re-use system is integrated into the urban environment.
9. The people approve and support the locally managed wastewater management system.
10. It is financially feasible to operate and maintain the system.

1.3.7 Mixing and Matching Wastewater Management Systems and Technologies

A much wider range and choice of wastewater management systems and technologies exists today than just a decade ago. But to mix and match different but complementary wastewater management systems and technologies to create contextual fitness is easier said than done. Some structure can be achieved by using the framework of the six elements listed above. An overview of the most important options available today is provided in Table 1-1. We discuss and outline each of these options in Chapter 3.

This book argues for the need for a paradigm shift in the management of wastewater in developing countries. The knowledge and technology that can enable this shift have been piloted in many countries, but there is a gap between the current availability of innovative pilot systems and the promotion and financing of large-scale deployments and roll-out of these systems. Because we are in the middle of a period of change, neither a cookbook nor a guideline is required (or possible) here. What is required is an open discussion on sustainability, appropriateness, and system thinking, and a shift in focus from technologies to management systems.

1.3.8 Stories from the Field

Sustainability only can be understood in relation to a given setting, context, or location, so the focus should therefore be shifted from general approaches and technologies to specific management systems that fit into a given setting at a given time. The importance of using the specific local context and culture as the starting point—as the framework for assessment, design, and implementation, and as the basis for sustainability—cannot be stressed strongly enough. We therefore provide real-life stories and cases to illustrate, discuss, and reflect on actual, not theoretical, appropriateness and sustainability.

Table 1-1. Overview of Six Elements of Appropriate, Sustainable Technologies and Approaches

Scale of System	On-Site, Cluster, and/or Centralized Systems
1. Wastewater Collection	On-site source management for reduced flow and/or controlled input On-site collection Cluster simplified sewer
2. Wastewater Treatment	On-site: pit latrines; dry composting toilets; biogas digesters; septic tanks with seepage pit, drain field, constructed wetland, or sand filter; greywater reclamation units Cluster/centralized: ponds, trickling filters, sand filters, constructed wetlands, overland flow Combinations
3. Energy Consumption	Gravity-based systems Pumps powered by renewable energy Siphons
4. Urban Integration	Invisibility Multifunctionality Symbolic, aesthetic, or topographical integration
5. Re-Use and Re-Entry	Re-use: land application for agriculture, industry and business, housing, recreational or environment Land-based re-entry: subsurface seepage or surface sludge disposal
6. Organization and Finance	Local level: appropriate low investment and O&M costs; effective cost recovery; decentralized local organization National level: enabled through sustainable institutional, financial, and legal setup, and political will and stability

Chapters 4 through 11 present a number of true stories and actual cases from developing countries.

These case stories and context are specific, complex, and detailed, and are therefore hard to understand, let alone describe. We therefore begin each case with the story behind each system: how was it created, by whom, why, with what considerations, obstacles, and possibilities; who supported it and who was against it; and why it finally came to look like it did. Then we step back a little and reflect on the case from the larger perspective of appropriateness, sustainability, actual impact, local fitness, and robustness.

The best case stories are real tales from the field, told by the people who were directly involved. This has guided the selection of case stories here; they are cases that we or our partners were personally involved in. Because we mainly have been working intensively on wastewater management in southeast Asia, most of the case stories are from that region. This does not mean that the cases are not relevant for other regions. We have been working in many different countries all over the world for the last several decades, and have found that the stories and lessons learned from the case stories in this book are of general relevance for working with wastewater management in all developing countries. Furthermore, throughout the book we have included small case reports from other developing regions.

Our other primary criterion for good case stories is that they can be told with depth, not necessarily that they are success stories. Design, implementation, and operation of wastewater management systems in developing countries is extremely challenging; more often than not, the system either fails or is poorly operated and functioning. We certainly do not want to pretend that our case stories are success stories. We report on cases from a certain point in time, knowing well that some installations might fail and some have perhaps already failed or are not performing satisfactorily. The key is that we, and others, will only learn and improve by informed, thorough, and in-depth stories from the field.

1.3.9 Smart Technologies

Because all wastewater management systems contain certain specific technologies, we close each case story with descriptions and reflections on the technologies used in the specific case. Linked to the need for alternative wastewater management systems in developing countries is the corresponding need for new technologies. Some technologies point toward the future, some more toward the past. We have in each of the cases described and discussed promising, potent technologies we think might have future potential.

We have learned that potential future technologies have both rational and aesthetic elements. They are effective, simple, light, moveable, low-energy-demanding, user-friendly, intelligent, interactive, and beautiful. They may not have all of these features at once, but they will have some of these characteristics. The 10 technologies we will highlight and discuss in the book are:

1. Septic tanks combined with subsurface irrigation (Chapter 4, Section 4.3.1)
2. The technology of landscaping (Chapter 5, Section 5.3.1)
3. Vertical subsurface-flow constructed wetlands (Chapter 6, Section 6.2.1)
4. Urban integration of wastewater management systems (Chapter 6, Section 6.2.2)
5. Siphons (Chapter 6, Section 6.2.3)
6. Separate wastewater collection systems (Chapter 6, Section 6.2.4)
7. Solar-powered pumps (Chapter 6, Section 6.2.5)
8. Rotating biological contactors (aero wheels) (Chapter 7, Section 7.3.1)
9. Horizontal subsurface-flow constructed wetlands (Chapter 8, Section 8.3.1)

10. Combined pond and surface-flow constructed wetlands (Chapter 9, Section 9.3.1).

This book hopefully will start a discussion among professionals about the future of wastewater management, and about the need for a paradigm shift in wastewater management in developing countries. We call for a new agenda at universities and among professionals in the field for the benefit for the billions of people currently living without adequate sanitation.

2

Reflections on Sustainable Wastewater Management

⟫⟫⟫ 2.1 Discussing Appropriateness and Sustainability

The basic characteristic of an appropriate wastewater management system is that, at a given time in history, it fits well with the local setting and culture for which it was developed, thereby ensuring its relevance and sustainability. Everyone would probably agree with this but, when we move beyond the nice phrasings into the real world, much disagreement exists within the field of wastewater management.

Some wastewater management systems do fit better into a given setting at a given time than others do, but which systems? The team behind this book, given their collective years of experience with wastewater management in different locations, from different angles, and from different professions, discussed and brainstormed our lessons learned on these basic questions. What had gone wrong and what right on certain projects? Why did existing technologies so seldom fit in? Why did it seem like the local contextual assessment so often failed? Why was it that each one of us had certain "black holes" (often not small ones)—knowledge, sometimes very basic, we did not have, and information we had been given but sometimes forgot because it was not intuitive or part of our own contextual, nonreflexive knowledge. We quickly settled on our first important conclusion:

> Design, implementation, and operation of appropriate wastewater management systems requires input from teams of experienced people with different backgrounds: practitioners and scientists; (inter)nationals and locals; and technicians and administrators.

Then we went on to discuss which issues we had found to be fundamental to developing successful wastewater management systems in developing countries, and in Thailand in particular because this was where our discussion took place. Some of the issues from our discussions are provided and reflected upon below. All of them are obvious in the sense of Yes, of course they must be included when we design and implement wastewater management systems, but it is our experience that they are often overlooked, by internationals, nationals, locals, and by us. Some of these issues we know but, to be honest, we tend to forget them time and again when designing, implementing, operating, and rehabilitating wastewater management systems. Some issues we knew but chose to overlook, often with dire consequences. Some of us honestly did not know about some issues. All of these issues influence the creation of appropriate and sustainable systems—what works and what does not. All have influence on getting the context right. The following eleven issues, listed below in no particular order, are therefore in many ways our personal lessons learned as a team during the last decades. These fundamental issues are the basis for the definition of appropriate and sustainable wastewater management as discussed in Chapter 3.

⟩⟩⟩ 2.2 The Eleven Fundamental Issues

Issue 1: We tend to forget the basics of temperature and climate and their influence on appropriate and sustainable wastewater management systems.
Developing countries in tropical climates are normally gifted with rich amounts of year-round sunshine and more or less constant air temperatures around 30 °C (86 °F). This provides optimal conditions for wastewater treatment processes because efficiency in biological wastewater treatment peaks at around 30 °C to 40 °C (86 °F to 104 °F) (Fig. 2-1).

Wet and dry seasons, evaporation rates, and high and fluctuating rain intensity are also important factors when designing and operating wastewater

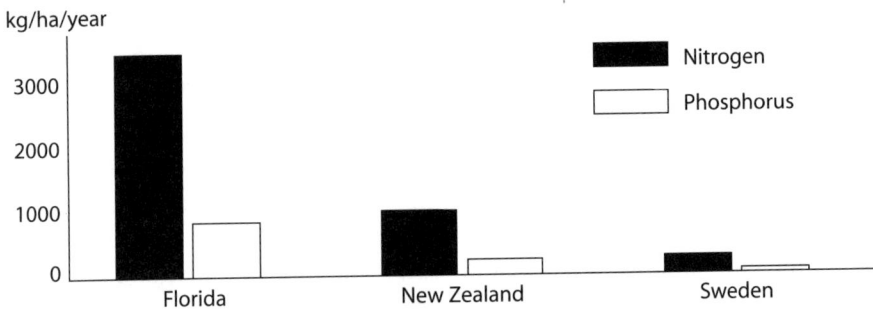

Figure 2-1. Uptake of nutrients in constructed wetlands in different countries. The wetlands in warm and sunny Florida are up to 15 times more efficient than in more temperate countries like New Zealand and Sweden.

Source: Adapted from Fujita Research (1998).

management systems. Most storms in the tropics are characterized as heavy showers, with very intense shock loadings of rainwater. In Bangkok, for example, more than 80% of the annual precipitation falls within 100 hours, and more than 200 mm (8 in.) of rain can fall on the city within 24 hours.

How can high evaporation rates be utilized? What do the very strong rainfalls mean for the collection system? What consequences does a long dry period have? The consequences of ignoring these important climatic factors can be illustrated by "first flush" problems and soil erosion.

First flushes (the first strong rain after a dry period) are typical tropical phenomena. Their environmental impact is immense because they flush vast amounts of heavily polluted sludge in the sewers, which accumulated during the long dry period, into the rivers or the coastal area. If wastewater management systems are to prevent adverse environmental impacts on coastal ecosystems, they must manage the impact of first flushes. Regular cleaning of the drains or sewers in the dry period would be one obvious solution.

Soil erosion is another issue aggravated by tropical climates. The heavy rainfalls result in vast amounts of soil being flushed down the hills into the combined drainage and sewer systems. Fast-developing areas—often tourist areas like Koh Samui and Phuket in Thailand—have immense problems with these huge amounts of soil choking the drainage system and the resulting malfunctioning of the wastewater collection and treatment system. The combination of high rainfall, high development rates, and hilly topography creates enormous problems for the drainage and wastewater collection systems—problems that must be solved before wastewater treatment systems even can be considered.

Sustainable wastewater management systems must be able to deal with these specific tropical climatic issues. They must include solutions for, among other things, (1) the problems occurring in the rain-free periods: the increased risk of waterborne diseases, odor, stagnant waters, and visually very poor water environments; (2) the problems occurring in the shift from the dry to the wet period—first flushes create a shock loading of black wastewater because all accumulated matter in the drainage system is pushed downstream to the major outlet or treatment facility; and (3) for the problems occurring in the rainy seasons—flooding and overflow of wastewater reaching the street level.

Issue 2: We tend to ignore the already implemented, on-site wastewater management systems.

Wastewater management planning and systems should always take their starting point in existing households that have already taken care of their wastewater problems (of course, with varying degrees of efficiency and success). Many on-site systems such as septic tanks and seepage systems have been constructed, are in operation, and have already been paid for through private investments.

To restate the typical situation in most developing countries: (1) black wastewater from toilets is separated from greywater (kitchen and baths) at the household level; (2) black wastewater is treated by septic tanks or similar systems within a single plot before it is discharged to the soil through local seepage systems, or, alternatively, by overflow to a drainage system; (3) greywater is seeped or discharged directly (without prior treatment) to public stormwater drains, but does sometimes pass through a septic tank before being discharged.

This on-site system relies on both the obvious need of each household to solve its wastewater management problems and on governmental housing rules and regulations. Investments are distributed among private landowners and do not rely on governmental funding. Because the implementation, legislation, and enforcement of on-site household wastewater management has in general proven successful, such policies have often been extended. This means that almost all urban household estates, high-rise buildings, institutions, commercial complexes, and industrial sites have local wastewater treatment facilities within their premises. As a result, the vast majority of wastewater is pretreated before it reaches the public drainage system.

These private on-site wastewater management systems, and their corresponding private investments (large, when accumulated), are often ignored when public centralized systems are planned and implemented. This normally results in conflicting coexisting systems installed simultaneously in the same areas. For example, the presence of on-site treatment facilities can actually reduce the efficiency of an advanced wastewater treatment facility because they result in a low level of organic matter in the wastewater reaching the treatment facility. Or they can create conflicts of interest regarding legal and financial requirements: "We have already invested as required; why should we pay again for being connected to the public system? Where is the law that can force us to connect? Why do we need a centralized system if distributed technologies already are in place?"

Issue 3: We tend to ignore the importance of managing sludge from septic tanks.

Where can the best environmental impact for the least resource input be achieved? Certainly, one important area is more efficient sludge management. Where on-site wastewater management systems have been installed, sludge must be removed regularly for such systems to have a cumulative positive impact. This is seldom the case. Sludge is removed when the septic tank gets blocked up or when the house owners, at night during a heavy rain, pump and empty their septic tanks into the nearest drainage system—this is regarded as the easiest, cheapest, and least bothersome way of getting rid of septic tank sludge and wastewater! But not only must sludge be regularly and correctly removed, it also must be discharged in ways not detrimental to public health and the environment. There is still a long way to go from the traditional dumping into the nearest stream to appropriate and sustainable strategies for disposal and re-use, such as biogas production, soil application, and other ways of re-using sludge.

Access to on-site treatment facilities is often difficult because many septic tanks are built under the house, in the backyard, or are hidden under concrete or pavement, which complicates regular and effective emptying of septic tanks and challenges effective maintenance. Not many professionals or administrators want to deal with the issue of sludge handling (a low-status area within a low-status sector!) but, if the focus is to be on fast, important, and positive impacts on ecosystems, this is probably the most efficient starting point.

Issue 4: We tend to forget that it is not possible to force people to connect to public drains.

One of the multiple reasons for the failure of large-capacity, centralized, and advanced wastewater treatment plants is the absence of legislation forcing private landowners to connect to a public sewer or drainage line (and even if such laws exist, lack of enforcement takes over). The nonexistence of laws forcing connection to sewers leads in many places to very low connection rates, which again means that the actual loading rates at centralized municipal wastewater treatment facilities vary greatly from the volume predicted by population data. The wastewater from households does not get into the sewers. All existing households may have already installed on-site systems, but when centralized systems are introduced the private households are expected to pay for all expenses of construction work, excavation, and tearing up floors, bathrooms, or parking lots to install new pipes for the connection and redirection of grey and black wastewater flows from the private plot to the public collection system. Understandably, relatively few households make the effort to connect to such centralized systems.

Numerous examples can be found of large, expensive centralized wastewater management systems being implemented that comprise only a main and secondary sewer, but no tertiary sewer lines linking the system to the individual households. It is somehow expected that the connection to individual houses will take place automatically, or the decision is postponed to some time in the future.

Historically, urban sewage systems have been installed in a fashion similar to that of urban water supply. Infrastructure development starts with the trunk sewage system and wastewater treatment facilities, followed by the sewer network in each community. Once these public components have been installed, residents can connect their homes to the sewer network. In many developing countries, this approach has been markedly less successful for

> **Box 2-1. Waterborne Sewage Systems in Africa**
>
> In 1973 a full waterborne sewage system was installed by the Ghana Water Sewage Corporation in central Accra with World Bank assistance, covering 1,000 hectares and involving 28.5 km of sewers. This effort is a classic example of services unaffordable by the prospective beneficiaries. The system never worked, partly because narrow and crooked streets and below-standard housing and plumbing hampered connections to the system. Only 6.5% of the available connections were utilized. In this as in many other examples, the supply-driven approach to sanitation system wasted immense investments. Inappropriate designs, neglect of user requirements, inadequate maintenance, and ill-equipped operating agencies created a continuous drain on government resources and a disincentive to governments and donors contemplating further sector investment. Users became disillusioned when the promised improvements failed to materialize; they refused to pay for inadequate services, leading to further deterioration of the system (UNEP 2002c).

sanitation than for water supply. Sewer systems installed using this approach have often been highly underutilized, as in Bangkok, Accra (in Ghana), and Mumbai (Bombay). In other cases, plans to install citywide sewer systems are never implemented due to prohibitively high costs.

Issue 5: We tend to ignore the fact that centralized collection systems are stormwater drains carrying mainly rainwater and greywater.
Rural areas have, in general, no wastewater collection systems and therefore rely solely on on-site treatment systems. Wastewater management in developing countries is predominantly an urban issue. The typical picture is that in the early stages of urbanization, old irrigation systems are converted to serve as storm drains, and the water management is focused on flood protection rather than management of domestic effluent from the households. Drainage systems exist in almost all cities and they are mostly designed as gutters, canals, trench boxes, or drainage pipes along the streets.

As the urban areas densify, more and more domestic wastewater finds its way into the drainage systems and the drainage systems will, over time, transform into combined open or covered stormwater canals and sewers—sewers in periods with no rain and combined systems during and after rain (Fig. 2-2).

The drainage systems of Accra, Lagos (Nigeria), Dar es Salaam (Tanzania), or Addis Ababa (Ethiopia) typify the mixed nature of wastewater in most big cities in developing countries. Formal and informal drains are filled with stormwater, septage, greywater, and solid waste. In the rainy season drains flood and overflow, and in the dry seasons they become informal waste dumps or

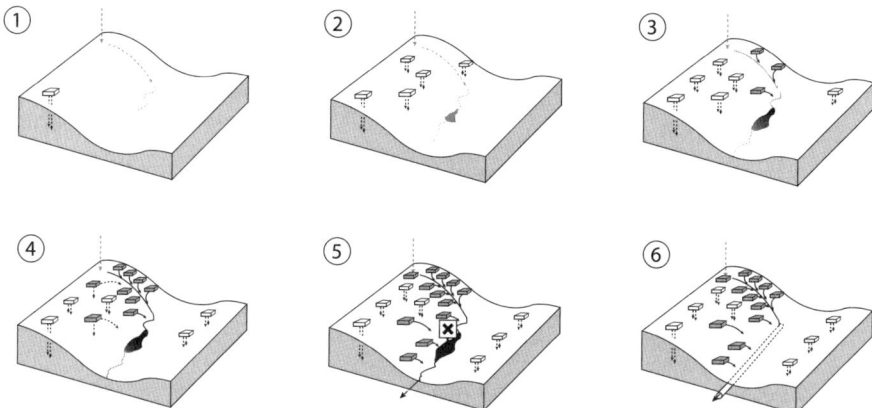

Figure 2-2. Elements of typical evolution of sewer systems. (1) Dispersed human settlement. Stormwater seeps into the ground or evaporates. (2) Seepage and run-off. Increased groundwater level causes temporary floods in the lowlands. (3) Few houses discharge to a natural ditch. Rising water flow and slight pollution of surface water. (4) Densification forces some houses with seepage to discharge to the ditch. Problems with smell and visually dirty drains. (5) Numerous dwellings discharging to the drainage system, causing polluted open waters and public health issues. (6) Covering or piping the drains. Connected households coexist with on-site treatment and seepage households.

stagnant cesspools. In medium- to high-density residential areas of most African cities, open storm drains are common and in many cases act as open sewers, particularly for the conveyance of greywater and overflow from septic tanks. In addition, most industrial wastewater is discharged into these same drains. This practice changes the characteristics of wastewater drastically (UNEP 2002c).

Wastewater collection systems in urban areas are therefore basically old stormwater drains that carry stormwater mixed with greywater and some effluent from septic tanks. The resulting pollution levels of the water in these collection systems are therefore generally much lower than those seen in developed countries, where all wastewater (black and grey) is discharged directly to closed-loop collection systems.

The gradual conversion from stormwater to combined collection systems has another implication. Open stormwater systems are normally designed as gravity drains with a low gradient of about 1% to 2%. Such a low gradient is insufficient for effective collection of wastewater. Furthermore, if there is no regular leakage maintenance or cleaning, only part of the wastewater will reach the treatment plants because the leaking, broken, and clogged drainage/sewers will cause exfiltration, overflows, and loss of collected wastewater.

It is also important to note that different urban or suburban areas (often close to each other) are often at different stages in this development. A study of nine adjacent districts in suburban Bangkok showed that all of the six development stages illustrated in Fig. 2-2 coexisted; this emphasized the importance of assessing the urbanization stage of a location before wastewater management systems are designed and implemented (Laugesen et al. 2004). Many examples exist of centralized systems that have been introduced in urban area development Stage 2 and Stage 3 locations, leading to very low connection rates and very little wastewater to treat at the treatment plant.

Although planners normally assume a sequence of land acquisition, planning of infrastructure, and then construction and building, the reality in developing countries is more chaotic, as illustrated Fig. 2-3.

Issue 6: We tend to be unaware of the complexity of the existing collection systems.

Surveys of existing combined drainage and wastewater collection systems in Thailand have shown a high degree of system complexity, between cities and also between different parts of the same city (Laugesen et al. 2003). These surveys found stretches of collection systems with no water; some gravitating the wrong way; some carrying only rainwater; some with few household connections, some with many, and some with none; some with many leaks, some with few; some functioning only as mains with no secondary lines connected; some connected to other lines in rather unpredictable ways; and some carrying large amounts of black and grey wastewater (which they should not, but did) while others carry only greywater.

Often, the shape of urban cities in developing countries does not match the requirements for conventional sewers to be laid down. This is certainly

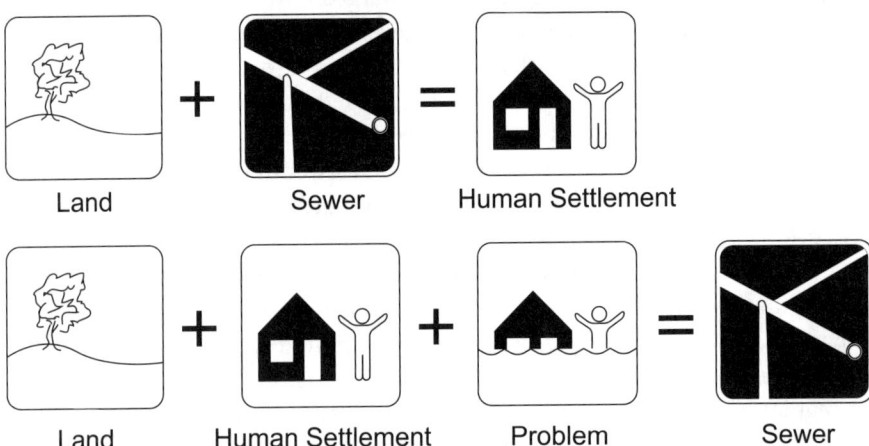

Figure 2-3. Logistics of sanitary infrastructure. The top equation is the present-day typical European model. The bottom equation is the present-day typical model in developing countries.

true for the shapes of slum areas and not-so-organized suburban settlements. It is not unusual in these areas to see streets that are too crooked or narrow for the required standard design codes for pipe excavation, manhole construction, and covers.

Issue 7: We tend to overdimension treatment plants compared to actual loading rates.

The low level of connection rates, the lack of laws to force sewer connections, the fact that sewers are often drainage systems that mainly carry greywater diluted with rainwater, and the complexity of the existing drainage systems are key characteristics of wastewater collection systems in developing countries. These characteristics often lead to overdimensioned, centralized treatment systems. The gap between the design capacity of the advanced treatment facility and the actual loading rates at the inlet to the plant is often considerable and often leads to scrapped or inefficient treatment facilities, unused equipment, treatment units closed to reduce energy costs, or malfunctioning treatment processes.

Because the influent at wastewater treatment plants generally consists of pretreated black wastewater, slightly contaminated greywater, stormwater run-off, or even cleaner groundwater entering poorly installed sewers, existing advanced treatment plants are often oversized and/or unnecessarily equipped to function with high organic loading rates. Activated sludge treatment plants and similar facilities with expensive operation and maintenance (O&M) requirements have in many instances been unnecessarily established at outlets with low influent concentrations. Planners tend to ignore that incoming biological oxygen demand (BOD) levels of 40 to 80 mg/L would be normal in developing countries, compared to 200 to 300 mg/L in many developed countries.

Also, planners tend to ignore the fact that much larger contents of oil and grease can be found in wastewater in tropical developing countries compared to wastewater in developed countries. Cooking in many tropical countries involves large amounts of cooking oils and sauces, which naturally influence the ratio of oils and greasy waste flushed to the drains from kitchens and restaurants. Other sources also contribute to the high oil and grease content in the wastewater, and ignoring this fact often has severe consequences for the efficiency of the wastewater collection and treatment system. Drainage pipes clog, pumps choke, and air blowers block as oil and grease get stuck on pipes, pumps, instruments, and air distribution systems.

Issue 8: We tend to ignore the gap between centralized planning and local operation.

Many municipal treatment facilities have been financed and implemented by a central administration, with relatively little cooperation between the administrators and the technical staff in the local authorities. Typically, a top-down scheme is financed by the central government, implemented by a contractor as a turn-key project, and then is meant to be handed over to and be operated and maintained by the local authority.

The first two parts, financing and construction, are often a win-win situation for all involved decision makers, but when it comes to the last part, O&M, it often becomes more difficult. This last link often fails because the local municipality has neither the motivation, the knowledge, nor the operational finances to carry on the task. This is one of the many interlinked reasons why the majority of labor- and energy-intensive treatment facilities become malfunctioning. When the local municipality discovers its often prohibitively expensive financial obligations to operate and maintain the treatment facilities, they immediately start looking for cost savings: pumps are turned off, wastewater by-passes the treatment facility, and aerators are shut down.

Financial constraints, however, might not be the only problem for local authorities. Advanced treatment facilities need technically competent engineers, but the number of engineers qualified to operate advanced wastewater treatment plants is very limited throughout developing countries—especially in more remote areas. Furthermore, the pool of competent engineers who might find it attractive to work at a municipal wastewater treatment facility with uncompetitive salaries and uncertain budgets is even smaller. Wastewater management is low on the wish-list of most engineers, somewhat behind other engineering disciplines. A local authority can quickly find itself in a situation where it does not have enough skilled staff to operate the treatment facility.

Issue 9: We tend to find it hard to accept that some centralized systems have no logic, except for the financials that drove the implementation.
Wastewater management systems are implemented to improve public health, protect coastal ecosystems, or support the local business environment. Why are systems implemented first in some locations, and later in others? Generally because the basics—public health, the ecosystem, or the business environment—are relatively more sensitive in these locations compared to others.

What happens if these fundamentals are not addressed in the planning and implementation of wastewater treatment facilities? If the basic justification for a specific treatment system is lacking (e.g., there is no significant public health or environmental problem), malfunctioning and unsustainable systems are usually the result. Similarly, if a facility has been located in an area where the issues of wastewater, public health, environment, and so forth are relatively less important than in other nearby areas (e.g., if treated wastewater will flow directly into very polluted water downstream), problems of sustainability will also most often be encountered. Motivation will be lacking.

In a technical, rational world it would be expected that treatment facilities are built to solve specific and significant problems, and that the problems in a country or province have been prioritized so the most serious problems are taken first. Unfortunately, such technical rationality for decision making is rare. Selection and prioritizing are based on many different factors, among them, of course, political and economic reasons and motivations.

Thus, it is often difficult to justify the construction of a specific treatment facility at such-and-such specific location because the "value (health/coastal ecosystems) for money" equation is often rather difficult to see. Treatment facilities are often built at locations where wastewater is not really an issue compared to many other places. They are overdimensioned; they are unnecessarily advanced. They are, in essence, built with the main purpose of spending money—as much as the central budget or donor can and will make available.

Such facilities are primarily financial win-wins for the involved political and administrative decision makers. And here centralized systems have big advantages over decentralized systems. They are more expensive, easier to plan, and can be implemented in one initiative, reducing the time span of implementation and the number of decision makers to be involved.

> ###))) Box 2-2. Not Making Sense
>
> Systems lacking basic justification are by no means exceptions. Taking Thailand as an example, a rough estimate is that at least three-quarters of the implemented centralized treatment facilities would be hard to justify from a national perspective of rational wastewater planning and impact. Two examples will illustrate this (Fig. 2-4).
>
> Si Racha is a fairly large industrial port city with about 60,000 inhabitants, where all rainwater and greywater is collected by gravity to an outlet on the shore. Central authorities designed and constructed a pumping station to lift the water (BOD levels between 10 and 70 mg/L) backward up to a new advanced treatment facility located 5 m above sea level. Because the treatment plant was located in the city center with limited land availability, it was designed to be three stories tall. This required that all combined stormwater and wastewater collected from the city had to be pumped 15 to 20 m upward before treatment could take place.
>
> Because the electricity expenses for pumping came to weigh very heavily on the overall budget for the municipality, it was tempting for the local government to let the mixed stormwater and greywater continue its natural flow into the sea, thereby saving facility O&M costs. And this, of course, is what has happened. The treatment plant has discreetly become nonoperational. The point is that the treatment facility does not in any way improve the conditions for the citizens in the municipality: dirty wastewater still flows in and under the streets; residents must bear the burden of

costs and smell from the collection and treatment system; and the outfall is located in a nontourist area that does not affect sensitive or threatened coastal ecosystems.

Chum Saeng is a small, rural town on the banks of a river in the middle of the country. The river is large with a rapid flow feeding into an even larger river, which, after passing through polluted Bangkok, ends up in the Gulf of Thailand. All rainwater and greywater with average BOD levels around 50 mg/L is collected at outlets at the riverbank. However, the wastewater treatment plant has been located uphill behind the town, so that all combined rainwater and greywater must be pumped twice to reach the pond treatment facility. The municipality has therefore discreetly turned off the pumps, leaving only rainwater to fill the treatment ponds. Coercing the municipality to do otherwise would either require very strict enforcement by the central environmental authority located several hundred kilometers away, or a very eco-friendly mayor who would bear the cost of invisible and unmeasurable improvement of the environment for the municipalities *downstream*, and acknowledging that there would be no gains for his or her own municipality. The public health and environment in his or her own municipality is the same whether or not the treatment facility functions.

Mayors, politicians, or administrators who choose not to operate wastewater treatment facilities are often labeled irresponsible, careless, antienvironmentalist, or even corrupt, but sometimes the perspective of "not making sense" puts the issue in a different light. Municipalities in developing as well as developed countries normally have very limited budgets, and careful considerations and prioritizations constantly must be made to fulfill the many needs of the local population.

Issue 10: We tend to underestimate the problems with rehabilitating existing systems.

The reasons for the high number of malfunctioning treatment facilities include: the systems lack overall justifications; the systems require local operation of facilities that were centrally initiated and implemented; the facilities are overdimensioned, too advanced, and have highly complex collection systems; the facilities have low connection and loading rates; and the systems lack

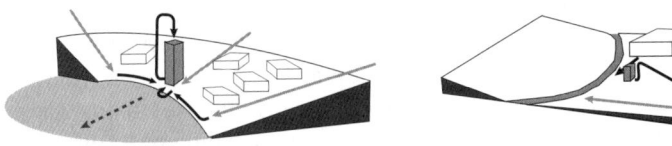

Figure 2-4. Si Racha (*left*) and Chum Saeng (*right*).

tropical fitness. These same reasons cause substantial problems in attempts to improve the operation of existing centralized systems, and we must be very careful when rehabilitating or improving such often failed or malfunctioning systems.

Rehabilitating existing centralized facilities is very difficult due to the fundamental problems of justification, finance, and O&M skills. Local motivation for rehabilitation becomes much more than a matter of technical upgrading of skills and repair of broken infrastructure or missing equipment—the challenges are deep and interlinked institutional, historical, motivational, and financial problems combined with inappropriate technology. Such problems most often cannot be solved, at least not presently, where other pressing issues exist and where the priorities in a given location do not support the required focus on, and substantial allocation of energy, resources, and competence to, centralized wastewater treatment.

Issue 11: We tend to plan and design based on very uncertain and unreliable baseline data and future projections.
Many traditional forms of data influence the planning for wastewater management systems. Planners need to know (1) the present and future numbers of inhabitants; (2) the number of connections required; (3) average water consumption rates; (4) the water/wastewater rate; (5) the black/grey waste-

---

))) Box 2-3. How Many People? How Much Water?

In a wastewater management planning exercise for a municipality in the outskirts of Bangkok, a design team needed to determine the number of inhabitants. The municipality had 21,435 registered inhabitants. However, an analysis of satellite images—counting the number of different types of residential buildings, multiplied with the average number of inhabitants per type—estimated that 55,000 people lived in the municipality. A follow-up house-to-house survey resulted in the number 52,000. The mayor, unofficially, thought that approximately 45,000 people lived in the municipality because this was the figure on which he based his election campaign. The team found that even the basic issue of how many people actually lived and produced wastewater in the municipality involved a huge possible margin of error. Basic data for water consumption were also problematic. It was not possible to get reliable consumption data from the public water company because that agency might not have collected reliable data or, if it did, it was not willing to share it (organizational competition with the wastewater authority, or data secrecy due to internal profit-sharing mechanisms). Whatever the reason, water consumption data were not available (Laugesen et al. 2004).

water rates; (6) the average leakage and infiltration rates; (7) the average loading rates; (8) the urban development rates; and (9) the present and future costs, just to mention the most important. From these data, planners can predict future needs and initiate the detailed design of appropriate wastewater collection and treatment facilities. However, anyone with practical working experience in developing countries will know that each of the above questions, which might appear simple and straightforward, in fact are very difficult to answer.

The collection of baseline data is not as simple as calling different agencies or looking in the official statistics and then proceeding from there. Collectively, wastewater planners end up with very unreliable baseline data, which is the data they use for predictions, system design, cost estimations, and finally for actual investments.

Besides baseline data, traditional wastewater planners need to estimate and predict future requirements, which leads to the next bundle of data problems. For example, consider projections of the rate of development. Many countries are developing rapidly, some with annual growth rates exceeding 8%, but growth rates are often very uneven between years and between regions and districts, and are therefore difficult to predict *locally*. Some areas with zero growth suddenly explode into 5% to 10% growth rates, and some areas experience rapid, often seasonal, tourism-based growth with thousands (or even millions) of visitors annually flocking around estuaries and beaches, creating a whole new set of problems for the wastewater management planner.

For this book, our team discussed and reflected upon several other "tendencies," but we will stop here. The above were a mix of some of the tendencies we found most important and which we often have encountered during our daily discussions and work with wastewater management planning and operations in developing countries. They bring us, through the back door, to the issues of appropriateness and sustainability, which is the topic of the next chapter.

3
Elements of Sustainable Wastewater Management

››› 3.1 Framing Appropriateness and Sustainability

Two main issues are important for planning appropriate and sustainable wastewater management systems in developing countries and elsewhere. First, the wastewater management system itself should include all of the following six elements (Fig. 3-1):

1. An efficient wastewater collection system
2. A sustainable wastewater treatment facility
3. Management of treated wastewater and sludge for re-use purposes
4. Reduction of energy consumption
5. Integration into the urban environment
6. A sustainable financial and organizational setup.

Included in a wastewater management system as a chain of interdependent subcomponents, these elements together create a closed-loop, cyclic wastewater management scheme. Not all of them can always be included but, if not, we should at least be able to answer honestly and sensibly why this or that element has not been included. Each element and multiple alternative varieties of each will be discussed and reflected upon in the following.

Second, it is important to use the specific context as the starting point, both in general and in relation to each of the above six elements. We consider the contextual understanding—the ability to read the site and specifically plan the most appropriate wastewater management system for the given area—to be the key underlying element in all planning of appropriate and sustainable wastewater management systems. To do this, one must assess the current state of the site in terms of population, flow rates, character of the wastewater, efficiency of existing or previous collection and treatment facilities, connection rates, laws and regulations, enforcement practices, local support, local habits, political preferences, incentives, and local lessons learned.

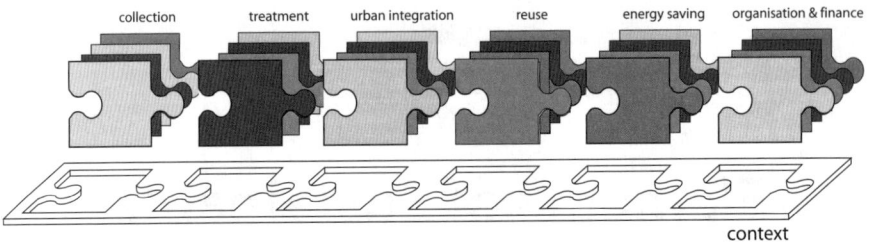

Figure 3-1. The six elements of appropriate wastewater management and the contextual fitness of each element.

The context is a large, complex, and diffuse cloud of multiple local parameters that together frame the project. Schematic models will never capture the complexity of the context, and it is our experience that presenting a universal checklist about how to properly assess the context never does the job. Getting the context right is not simple but hopefully this book will contribute to an understanding of this process and how to get it right (and wrong), as exemplified through specific cases and tales from the field (van Maanen 1988).

The approach proposed here, with its focus on context and cyclic systems, fundamentally counteracts the laziness of conventional planning, where the concepts of "scaling" and "copy and paste" seem to predominate. In contrast, we advocate site-specific responses that promote concepts such as utilization, optimization, adaptation, integration, and modification, as well as robustness, reliability, and compatibility.

》》 3.2 The Ten Nods of Appreciation

Early in the preparation of this book, we had a team discussion about when we had actually "nodded approvingly" during past on-site visits to established wastewater management systems, and all the times we had shaken our heads in disbelief. The latter has been covered in the "tendencies" discussed in Chapter 2, so now let us move on to the positivity of "nodding."

We nod approvingly when the planners and implementers get it right; when we are impressed; when we see something we had not thought of ourselves; when systems have been designed and implemented in an integrated manner; and when they comprise more than just treatment of wastewater. We nod when we see a cost-effective solution, a "good value for the money" system, and a way to solve the problem in the cheapest and most appropriate way. Nods are like benchmarks. So when do we nod? Here are 10 guiding (nodding) principles for appropriate and sustainable wastewater management.

▶▶▶ Box 3-1. Toward Sustainable Wastewater Management

In New Zealand during the last 100 years, wastewater management systems were conceived, built, and managed as if they were largely separate from the surrounding natural ecosystems. With the rapid increase in town sizes and better understanding of wastewater as a source of human disease, the concern was to transport wastewater away from settlements into rivers, streams, or the coast, where it was expected that dilution would take care of the problem. One effect of this strategy was to concentrate wastewater, thereby placing more pressure on the receiving ecosystem. This allowed some sectors of a community to forget or ignore the environmental effects and to see management of wastewater as independent of natural systems. Issues such as soil types and water tables were irrelevant because the system by-passed the natural process of wastewater management. But the rivers and coastal areas were eventually overwhelmed by the volumes of wastewater they were expected to handle.

From the 1950s onward, concern about effects on the ecosystem and on amenities and recreation forced the active treatment of wastewater. In the early years, this was mainly for health reasons, but later on it expanded to include treatment to a level that would minimize the adverse impacts on the receiving waters. The wastewater system still by-passed natural land-based percolation into soils, but it had been partly reconnected to the natural system by a minimum requirement to think about effects.

In recent years, the emerging view has been that wastewater systems should be integrated into natural processes. Of course, the new so-called *ecosystem-focused* or *integrated wastewater management approach* is not new. Many smaller communities and some farms and businesses use on-site systems that closely fit this kind of approach. In terms of designing the technical solutions for wastewater systems, there is now a shift from the conventional, linear, end-of-pipe technology to integrated water and wastewater systems.

The 2002 New Zealand Waste Strategy took significant steps to change the way wastewater is regarded. A major focus was on creating a circular process, which involves re-use, rather than a linear process from use to disposal. The result has been the addition of a re-entry management element to the collection and treatment parts of the system. Also in recent years, the costs of wastewater systems have sent some communities looking for ways to reduce that burden. This has resulted in thinking about the front end—the management of wastewater at the source—and the reclamation of treated wastewater to provide a re-useable water source (ME/NZ 2003).

Guiding Principle 1. On-site wastewater systems are preferred.
This implies that, whenever possible, wastewater should be managed (produced, treated, and re-used) on-site. Wastewater should be treated on-site to a level making it suitable for infiltration through the natural soil matrix and thus be recharged back to the water cycle through groundwater sources. Domestic wastewater is rarely a major problem if it is not collected, accumulated, and discharged at a single outlet. Wastewater is managed best when it is most invisible: no open drains, no big sewers, no large, high-tech treatment facilities—only effective small-scale, low-cost on-site systems, paid for and managed by each wastewater producer, either individually or collectively.

Guiding Principle 2. Short, gravity-based, separated wastewater collection systems are preferred.
This implies that 1 m of collection system is better than 10, 10 is better than 100, and so on. The shorter the distance wastewater is transported from source to treatment and re-use, the better. This also implies that transport by gravity is preferred to pumping stations and rising mains, and pumping wastewater once is preferred to pumping it twice. Finally, this implies that transport of domestic wastewater should be separated from transport of stormwater run-off and industrial wastewater, and that wastewater should not travel in the open. Closed pipes are preferable for conveyance of sanitary wastewater, as this reduces problems related to health, environment, and smell. Conversely, stormwater should be managed within the local urban landscape, [i.e., following the principles of sustainable urban drainage systems (SUDS)]. Shorter distances and separated wastewater and rainwater also results in smaller and cheaper wastewater collection systems because the diameter of the pipes can be reduced considerably, as well as better treatment efficiency because wastewater can best be treated when the concentration of polluting matter is high. Adding stormwater dilutes the wastewater, making treatment more expensive and less efficient.

Guiding Principle 3. Optimized household connection rates and source control are preferred.
This implies that all required household connections to the collection system are established and that control of what enters the collection system is in place. Have all or most households been connected? Especially in locations where no regulation or custom exists, how can these households be forced to connect? Have laws been enacted and are they enforced? Have oil and grease traps been installed to keep those materials out of the system? Have non-domestic pollutants or black wastewater (not pretreated) been excluded or dealt with? Have issues of first flush, soil erosion, and infiltration been considered and dealt with?

Guiding Principle 4. Simplicity, robustness, and local fitness of the wastewater treatment system are preferred.

This normally means that we are most impressed when systems have been developed that are easy to understand, construct, and maintain. Typically, this could be systems like ponds or constructed wetlands, or other systems that are based on natural processes and use as little mechanical equipment as possible but still reach the required treatment standards. We also nod approvingly when we see a good mix and match of treatment techniques, thereby making the system as robust as possible, and when an attempt has been made to fit the treatment facilities into the landscape, and when the treatment plant is pleasant to look at and visit.

Guiding Principle 5. Utilization of smart technologies is preferred.

Each component in wastewater collection, treatment, re-use, energy consumption, and so forth involves the use of specific technologies to transport, lift, purify, and distribute wastewater. Some technologies are more appropriate than others; some technologies are future-oriented whereas others are part of the past. Technologies that are part of the past are normally rather easy to recognize. Twenty years from now, will we still have to use 2-ton steel vehicles to transport our 70-kg bodies around town? Will we, at high cost, collect and treat polluted water (a valuable resource) just to discharge it directly into the ocean? Fifty years from now, will a wastewater treatment facility still look similar to a nuclear reactor? Predicting which technologies will not be part of the future is, however, much easier than predicting which technologies will be.

It is generally believed that new potential wastewater management technologies will be:
- *Effective, simple, light* (mobile, easy to install, easy to remove)
- *Robust* (reliable, durable)
- *Low-energy-demanding* (utilize renewable sources and natural processes: sun, wind, waves)
- *Low-cost, user-friendly* (easy to build, operate, maintain)
- *Intelligent and interactive* (self-adjustable, self-adaptable, upgradable)
- *Re-usable* (decomposable, recyclable, environmentally sensitive)
- *Beautiful* (aesthetically pleasing, exemplifying the difference between human appeal and purely mechanical requirements, clever)

The "smart technologies" might not be all of this at once, but they will have some of these characteristics as main features. Promising future technologies use the built-in knowledge of nature; they have few negative impacts; they are based on biological or intelligent technologies and they are often both rational and aesthetic. It is difficult to define smart technologies because they encompass different hard and soft values, but they are usually recognizable when encountered.

Guiding Principle 6. Re-use of the treated wastewater is preferred.
In general, treated wastewater should not be discharged directly into streams, rivers, or coastal ecosystems. Whenever possible, land applications should be integrated into the wastewater management system, and re-use of treated wastewater for gardening, agriculture, golf courses, soil conditioning, or for new and innovative purposes should be optimized to its fullest. The degree of re-use depends on, among other things, land availability and cultural sensitivity, but re-use will become an important component of all future wastewater management systems.

Guiding Principle 7. Low energy consumption is preferred.
The lowest energy consumption is achieved when the system is fully gravity-based, from households to re-entry. By utilizing local topography and optimized design, this is possible for at least some wastewater management systems. If a fully gravity-based system is not possible, energy consumption should be kept at a minimum by, for example, using as few and small pumps as possible, by utilizing siphons (e.g., in vertical-flow constructed wetlands), or by having energy supplied by solar power, wind power, biomass, waves, or other renewable energy sources.

Guiding Principle 8. Integration into the urban environment is preferred.
This implies that each component of the wastewater system must become an integrated part of the urban landscape and the community. Treatment facilities that consist of ugly concrete structures or have foul smells are located as far away from the urban landscape as possible, increasing operating and maintenance (O&M) costs. However, treatment facilities can be smart, beautiful, and useful when properly integrated into the local context. Underground collection pipes and underground pumping stations with odor reduction features can be integrated into urban functions such as parks, parking lots, green fields, and recreation areas; the options for urban integration are numerous and new innovative approaches are increasingly being applied.

Guiding Principle 9. It is preferred that the connected communities support the applied, locally managed wastewater system.
This implies that communities experience direct benefits of the locally managed wastewater system, such as improved wastewater disposal, public health, or a better beach and coastal environment. Whether the system becomes sustainable is determined by local involvement, commitment, and sufficient local technical, organizational, and managerial resources. Sustainability of wastewater management in developing countries cannot be guaranteed but it is a requirement that commitment, ownership, and professional human resources are taken into account in the planning, design, and implementation of wastewater management systems.

> **Box 3-2. Islands, Tourism, Wastewater, and Coastal Ecosystems: A Cocktail for Making Sense**
>
> The Caribbean region is mainly composed of small island developing states, many of which are major tourist destinations due to their attractive natural coastal environment. However, there is a real danger that inadequate action and investment in managing wastewater will harm the coastal ecosystems and the associated tourist attractions. The increased supply of potable water, together with improved living standards, concentration of the population on coastal belts, industrialization, and tourism have resulted in more and more wastewater to be disposed of. Considerable attention has therefore been paid to wastewater management in the last decade. A recent survey of wastewater management facilities in the region showed that the operational conditions in 61% of the 138 treatments facilities surveyed where labeled "good" or "moderate" (UNEP 2002a). This figure is high compared to other developing regions and suggests that the package of islands, tourism, coastal settlements, and nearby sensitive coastal ecosystems provides a possible cocktail for the development of sustainable wastewater management systems.

Guiding Principle 10. It must be financially feasible to operate and maintain the wastewater management system.

This implies that appropriate and sustainable systems should have recurrent O&M costs low enough that the local authorities are able to manage the system. This also implies that fees for connections, discharge, or use of treated wastewater should be returned for the operation of the system. For example, for privately owned on-site systems, O&M responsibility has traditionally been left in the hands of the householder. However, householder neglect has been a significant contributor to the problem of poorly performing on-site systems that eventually have to be replaced or upgraded to cluster or centralized systems. A promising alternative option is to adopt a centrally managed, fee-for-service maintenance program for on-site systems, thereby preventing such deteriorating performance.

3.3 Scale, Systems, and the Six Elements for Appropriateness

We have now defined the six elements for appropriateness of integrated wastewater management systems, highlighted the need for contextual fitness, and provided 10 guiding principles for appropriate and sustainable

> **Box 3-3. On-Site versus Cluster System**
>
> The wastewater management system on Waiheke Island, located in the Hauraki Gulf of New Zealand, traditionally consisted of conventional septic tanks and soakage field systems. However, because clay soils and difficult topography limited the use of this approach, in recent years a variety of alternative systems have been utilized, such as pretreatment via aerobic treatment plants or sand-filter units, and disposal via evapotranspiration beds or drip irrigation systems. These worked satisfactorily for lower-density residential development, but for the commercial center of Oneroa Village, with its high water-use activities, on-site systems became unsatisfactory and a full off-site reticulation and cluster treatment scheme was developed. This scheme determined that the most appropriate effluent discharge method was a constructed horizontal-flow wetland into an existing natural wetland. This system, commissioned in 2002, included a recirculating sand-filter system as a secondary treatment system prior to tertiary treatment in the constructed wetland. Sand-filter systems have a stable treatment process, low maintenance requirements, and the ability to accommodate large load fluctuations. A native tree- and shrub-planting program was implemented for the whole treatment plant site to provide beautification and visual screening (ME/NZ 2003).

wastewater management systems in developing countries. In this section we will look at choices and options for wastewater management systems and their six elements.

When deciding how to manage wastewater in developing countries, a distinction needs to be made between the management system and the technical engineering solutions that might be used within that system. Thus, there are the wastewater management *systems*, such as on-site, cluster, centralized, or a combination of these, which consider and deal with wastewater from source to re-entry; and then there are the specific wastewater collection, treatment, and disposal *technologies*, such as septic tanks, constructed wetlands, drip irrigation, or centralized, advanced facilities.

For any location, the most important thing is to choose the management system first. Choosing the technology comes second. A location, whether urban, suburban, or rural, often has a wide range of wastewater management options available. The decision depends on many contextual issues and characteristics, including the basic ones of local soil characteristics, groundwater tables, or proximity to estuaries, a coast, or coral reefs.

Any wastewater management system must deal with the issue of scale, whether single houses, a business, a farm, a group of sites, a whole community, or a city. The three general categories of scale are:

- *Individual on-site systems* refer to any system where wastewater produced on the site is treated and returned to the ecosystem within the boundaries of that site. This may be a hotel, a business, or a single home.
- *Cluster systems* are community systems for two or more dwellings. They are generally much smaller in scale than a centralized system. The wastewater from each cluster may be treated on-site by individual septic tanks before septic tank effluent is transported through a sewer system to a nearby location for further treatment, re-use, and ecosystem re-entry.
- *Centralized systems* refer to systems where all wastewater is collected at its sources and then transported through sewer pipes to a central facility for treatment. After treatment, the resulting effluent and sludge are discharged at a particular point, thus re-entering the ecosystem. As in the case of cluster systems, some treatment may occur on-site prior to the wastewater being transported to the central treatment site.

Variation is possible within these three levels of scale (Fig. 3-2). For example, a cluster framework can have some on-site pretreatment and the final treatment plant can be located off-site. Re-entry of wastewater can occur on-site or off-site.

The following provides brief descriptions of various options within these three levels of scale and the six system elements. The options include wastewater systems and technologies that are likely to be of interest to developing countries.

))) 3.4 Element 1: Wastewater Collection Systems

3.4.1 Management at the Source: Water and Pollutant Control

Regardless of the scale of the system, management at the source can substantially ease or reduce the cost of the collection, treatment, re-use, and ecosystem

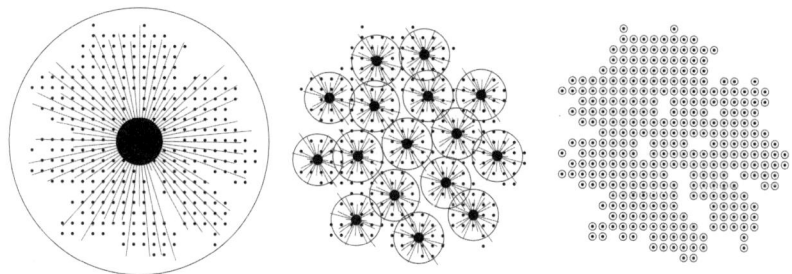

Figure 3-2. Principles of wastewater management. *Left to right*: central, cluster, and on-site.

re-entry. The *amount of water used and discharged* is a major factor in deciding on the type and size of a wastewater management system. Fairly obviously, water conservation (e.g., through improved toilet systems) can reduce the amount of wastewater that needs to be dealt with.

In areas with scarce water, such as on islands and in arid regions, water is too valuable a resource to be used as a basis for transport of human waste. In these locations on-site dry systems such as latrines, no-flush toilets with composting or incineration units, or flush toilets combined with centrifugal separators are preferable to water-based sanitation systems. This can also be an issue in areas lacking soil soakage capacity or experiencing high water tables. The predicted future water scarcity in many regions of the world further argues for non-water-based systems, like dry composting toilets (DCTs), being an important part of future wastewater management solutions. To use 50 to 80 L of high-quality drinking water every day to transport 1 to 1.5 kg of human waste to a wastewater treatment facility certainly will not continue to be appropriate.

The *types of pollutants* discharged are another factor in deciding on the type and size of a wastewater management system. The presence of toxic materials or heavy metals may demand a more technological level of treatment than would normally be used for domestic wastewater. The recent past has shown various good examples of management at the source of different types of pollutants. For example, some countries have now prohibited the use of phosphorous-containing detergents, so washing powder manufacturers have responded by replacing phosphorus with less harmful chemicals. As a result, wastewater phosphorous levels are lower and receiving waters better protected against rapid eutrophication.

At this time, management at the source is still seldom an integrated part of wastewater management systems, but it should increasingly be considered and implemented before launching into more advanced downstream systems. The basic principle is that appropriate wastewater management systems should always start by reducing the scale of the problem by reducing the volume of water and/or the scope and concentration of contaminants.

3.4.1.1 Collection System 1: On-Site

Wastewater treated and discharged on-site will normally be collected through simple pipes carrying wastewater to the treatment and/or land application system. For example, some on-site household drains consist of piping from the dwelling to a septic tank, and then effluent lines from the tank to soakage trenches. A case description of on-site collection is provided in Chapter 4.

3.4.1.2 Collection System 2: Conventional Off-Site Collection System

For wastewater treated off-site, wastewater needs to be collected and transported to the treatment plant, and several options exist. In the conventional collection system, households connect to large, below-street sewer lines, which are reticulated in straight lines between manholes that provide access at every change in direction (manholes are a significant proportion, typically around 15% to 20%, of the total wastewater collection system costs). Energy to transport the wastewater may come from pumping or a combination of pumping and gravity.

The total cost of operating a water-flushed toilet is often eight times that of a pit latrine. Sewer operation usually needs 50 L per capita per day just to keep the wastewater flowing, which is about the same as the average *total* water use for the poorest half of the urban population in developing countries. This means that water-based sewer systems are hardly adequate for at least the poorer half of the population in urban areas in Africa, and are in general not sustainable in most African cities (UNEP 2002c).

Where suitable soil conditions are present, nitrogen removal from septic tank systems compares favorably with water-based sanitation systems. Septic tanks almost always provide better phosphorus removal than does water-based sanitation in the absence of high-tech-designed treatment works. Because water-based systems normally do not have disinfection or maturation ponds, nitrogen, phosphorus, and pathogenic bacteria pass straight through the system into the river, lagoon, or the sea, creating serious water and environmental pollution in many developing countries. In case of system failure, which is frequent, full water-based systems may pose the most serious threat to the environment.

3.4.1.3 Collection System 3: Simplified Sewer, Alternative Off-Site Collection Systems

Conventional collection systems have, as mentioned earlier, a number of disadvantages. However, a number of alternative off-site collections technologies exist—technologies that improve collection rates, reduce infiltration, reduce odor problems, and provide better separation of wastewater and rainwater. *Settled sewage, small borehole, condominium,* and *effluent drainage servicing systems* are some of the alternatives applied with success in developing countries. These systems all include a toilet-flushing mechanism; an on-site storage/settlement unit (septic tank); a network of solids-free pipes designed to convey the liquid portion of the sewage to a central treatment and/or disposal point; a mechanism for removing sludge from the on-site containers; and a treatment and/or disposal facility (Fig. 3-3).

These simplified sewer systems provide an appropriate alternative for sewered pour-flush toilet and septic tank systems, and may often be the only feasible solution in urban areas with excessive housing densities where it is practically impossible to have individual family latrines due to space constraints, or where unserviced septic tanks would represent serious health

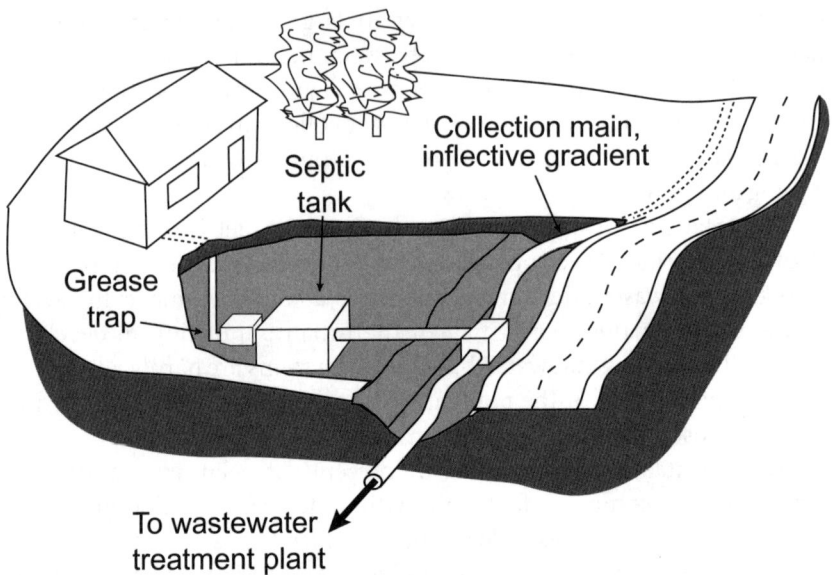

Figure 3-3. A simplified sewage system.

and environmental hazards. Examples of urban areas with these constraints abound in the suburban settlements of major towns and cities in developing countries: 30% of Addis Ababa, Ethiopia, large parts of greater Lagos, Nigeria, the Kibera slum in Nairobi, Kenya, and the squatter settlements of Gabo-

---

⟫⟫ Box 3-4. Nationwide Observations

A nationwide survey in New Zealand in 1995 identified 14 operating alternative wastewater collection schemes. The data indicate that in systems where the homeowner was responsible for septic tank maintenance, problems in the sewer lines frequently occurred, but that in all cases where local councils managed the total system, including on-site septic tanks, such problems did not occur. It was also found that treatment of the reticulated septic tank effluent was best achieved by oxidation ponds or wetlands. Mechanical aeration plants based on the mechanical activated sludge principle were not entirely satisfactory because the lower organic contents of septic tank effluent resulted in operating problems and poor performance. Overall, effluent drainage servicing scheme costs ran 12% to 45% lower than conventional wastewater collection costs, but such costing was very site-specific. Alternative wastewater collection offered particular advantages in locations with difficult topography and soil conditions—conditions that made conventional sewers expensive to implement (ME/NZ 2003).

rone, Botswana, fall into this category. Compared to conventional sewers, these alternative systems have smaller pipe diameters, flatter pipe gradients, shallower pipe depths, fewer access chambers, and no manholes, thus offering savings on capital, O&M costs, simpler design and easier construction, as well as simpler treatment requirements compared to the conventional sewage conveyance systems. In particular, they provide opportunities to retrofit sewer lines into unsewered smaller communities in difficult topography, or into high-density areas. It is estimated that these savings result in unit cost savings of 25% to 50% over conventional sewer systems in Africa, and savings in South African schemes in particular have been estimated at between 9% 43% (UNEP 2002c). This technology is widely known in Latin America but less well known in Africa and Asia.

The transport of solids-free wastewater is easier than the transport of wastewater containing solids. If solids are present, the sewer system should be designed with a sufficient slope to create sufficient water velocity to flush out the solids and prevent solids precipitation and pipe blockage. It might even be possible to apply inflective gradients as long as the overall hydraulic pressure in the system is sufficient. However, for this type of system timely desludging of the septic tanks is essential; otherwise, the sewer system will block. A major advantage of this system is that it may be better suited to existing conditions in developing countries because most households already have some kind of septic tank system that discharges into a drainage system or into gullies next to the roads.

Simplified sewers are sometimes termed *condominium sewers* in recognition of the fact that tertiary sewers are located in a private or semiprivate space within the boundaries of the condominium, and that the simplified sewer system includes these tertiary sewers, which often present the biggest problems for sewage in dense urban areas in developing countries. The designers and householders in the area to be serviced must determine which form or forms of condominium sewer will be most suitable for the local situation (Fig. 3-4).

))) 3.5 Element 2: Wastewater Treatment Systems

The three general wastewater treatment methodologies are on-site treatment, off-site cluster/centralized treatment, and treatment through a combination of on-site and off-site systems.

3.5.1 On-Site Treatment Systems

For on-site treatment, a range of treatment options are available, ranging from conventional septic tank and seepage systems to more advanced systems

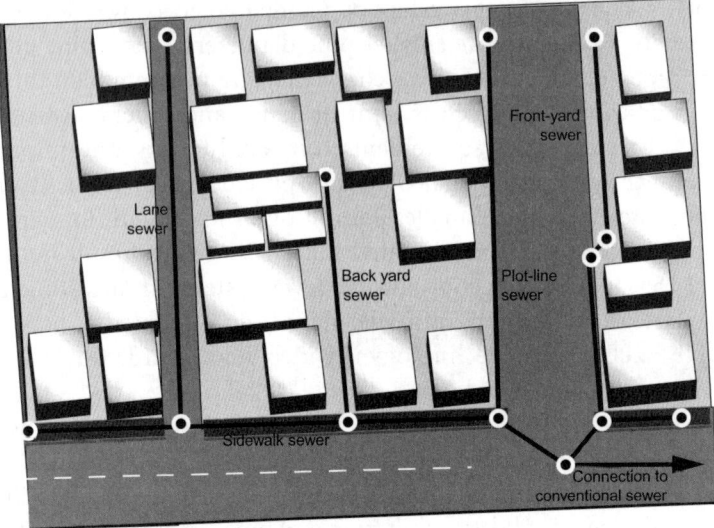

Figure 3-4. A condominium sewage system.
Source: Adapted from UNEP (2002c).

> **Box 3-5. Installation of Simplified Sewers**
>
> Simplified or condominium sewers in Brazil have been described as Latin America's most promising step toward the increase of sanitation coverage. This type of system was first used in 1982 in Natal, Brazil in a World Bank-funded project, and presently more than 4,000 km of condominium sewers have been implemented in that country—the largest installation of simplified sewers in the world. Simplified sewers have been successfully adopted into mainstream Brazilian sanitary engineering. Although most such schemes have been successful, some have failed mainly due to poor construction, poor institutional commitment, and poor maintenance. Average capital costs are about $22 to $34 USD per person. Simplified sewers have also been used in Bolivia, Colombia, Nicaragua, Paraguay, and Peru; in Asia since the mid-1980s (Sri Lanka has more than 20 schemes in operation); in Malang, Indonesia; and in Pakistan. In some parts of Africa, particularly South Africa, Botswana, Zimbabwe, Cote d'Ivoire, and Nigeria, interest in simplified sewers is increasing, particularly for alternatives that provide for flush toilets but have lower cost implications. Zambia introduced settled sewage systems in the 1950s in Lusaka, and has since extended the use of this system. In Nigeria, a sewered aqua privy in the town of Bussa opened in 1968, with the wastewater being treated in facultative waste stabilization ponds. Since 1989, South Africa has installed 21 such schemes, serving high-, medium-, and low-income communities (UNEP 2002c).

such as sand filters or constructed wetland systems. On-site systems service individual lots where all wastewater produced is treated on-site, and generally also re-enters the ecosystem on-site. The extent to which on-site treatment systems can appropriately be used is determined by the ability of the soils to absorb the treated wastewater; the characteristics of the local groundwater, including the level of the water table in different seasons; and the distance to sensitive ecosystems. Some soils are not suitable, whereas others may require a larger area for percolation. Sometimes underground water can be polluted by wastewater trickling through the soils, thereby preventing or limiting the use of on-site treatment systems.

On-site treatment systems are normally managed individually, often with suboptimal treatment results. However, an increasingly popular approach involves system monitoring and O&M inspections by a central agency to prolong the life of the on-site system while protecting the investment in the system's hardware. The cost of this centrally managed approach can, when prorated on an annual basis, equate to approximately the charge that would have to be collected in a cluster or centralized treatment system.

On-site treatment systems should always be supported by a fleet of pit or septic-tank-emptying vacuum trucks, together with public facilities for septage treatment.

Different on-site treatment systems may be appropriate for different contexts and for different types of wastewater. The following outlines some

---

))) Box 3-6. Survey of On-Site Treatment Systems

The coastal community of Manukau City, New Zealand, conducted an efficiency survey of on-site treatment systems in 2002. The community consists of about 280 dwellings, approximately 520 permanent residents, and some seasonally occupied holiday homes. Septic tank and soakage fields provide wastewater servicing for the majority of the properties. Environmental monitoring over several years had indicated fecal contamination of surface water drains, coastal waters, and local shellfish. The most likely contamination source was effluent from on-site treatment systems. More than 180 of the 280 sites were inspected and information was gathered to grade the performance of individual systems. The scoring system was based on assessment of environmental factors relating to soil conditions, soakage rates, groundwater level, and climate; and to site factors such as occupancy of dwelling, size of septic tank, maintenance frequency, and age of the system. The survey found that about half of all properties visited showed evidence of present or past failure. Failing systems were mostly located on slowly draining clay soils (ME/NZ 2003).

On-Site Treatment System 1: Pit Latrines

Conventional pit latrines, including pour-flush latrines and ventilated improved pit privies (VIPs, privies with exhaust chimneys drawing air away from the pits by convection or fan), are simple on-site systems that may be appropriate in locations where population density is low, groundwater level is low, the area is not prone to flooding, and where the community cannot afford a better system. The use of pit latrines is extremely common in rural areas or among the poor in developing countries because they are easy to operate and maintain, require no skilled labor for construction and maintenance, are low cost, and use no or very little water (as required for pour-flush latrines) for flushing. The basic principle is to hide human waste in deep pits ("drop and store"). The design life varies, depending on the number of users, but is normally from several years up to 10 years or more.

On-Site Treatment System 2: Composting Toilets

Composting toilets are especially appropriate for suburban and rural areas with lower-density population, in areas where the groundwater table is high, or where flooding is likely. Advantages include low initial investment, low O&M costs, no water requirement, no sewer network requirement, no pollution of groundwater, and production of valuable soil conditioner. Once full, the digestion chamber is left to compost over a period of weeks. During this time a second chamber is used. Finished compost is removed and may be dug into gardens or trenched around tree roots. Composting toilet systems require bulking material such as wood chips, dried leaves, coconut husks, or food waste.

The *urine-diversion toilet* adds to the composting toilet the separation of urine. These toilets are suitable for higher-density areas where beneficial

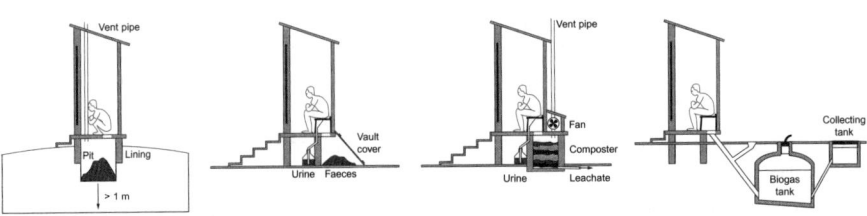

Figure 3-5. Conceptual drawings of (*left to right*) a pit latrine, a urine-diversion toilet, a composting toilet, and a biodigester.

Source: Adapted from Loetscher (1998), Shaw (1999), and UNEP (2006).

>>> **Box 3-7. Ecological Sanitation**

Ecological sanitation ("ecosan") is an approach to human waste disposal that aims at recycling nutrients back into the environment and into productive use. In the ecosan approach, human waste is considered a valuable resource. Until recently, re-use of human waste has been the norm in many societies such as in Europe and Japan and is still widely applied in rural communities in China and Vietnam, and in urban areas in Yemen, Mexico, China, and El Salvador (Sawyer 2001). The closed-loop ecosystem approach of ecosan builds on three basic principles. First, it promotes public health and prevents disease by treating human waste on-site rather than flushing it downstream for others to cope with. Second, it protects the environment while conserving resources. Finally, it recovers nutrients in human waste by returning them to productive uses and does not waste water as a valuable resource.

urine can be re-used as fertilizer or where groundwater pollution is a concern. Urine is collected separately from feces by a special design of the toilet bowl or pan; the toilet construction assists in the drying process of the feces.

On-Site Treatment System 3: Biogas Digesters

Biogas digesters are suitable for suburban and rural areas in hot climates, especially where households also have animal waste and where there is a need for gas for cooking. In a biogas digester, organic material is broken down under anaerobic conditions. This process produces methane that can be used for cooking and lighting. Biogas digesters operate best in warm climates because high temperatures ensure sufficient production of biogas and destruction of pathogens. The effluent from the digester may be used as a nutrient-rich fertilizer for agriculture and aquaculture, due to conservation of nitrogen during the anaerobic process. Biogas digesters may replace existing septic tanks by integrating the septic tanks as inlet chambers. The digesters use very little space; operational requirements are low; limited operator skill is required; desludging is only occasionally necessary (less than with septic tanks); they reduce energy costs; and they generate revenue by creating higher agricultural yields.

On-Site Treatment System 4: Septic Tanks Followed by Seepage Pits

Septic tanks followed by seepage pits are appropriate for areas with low to medium population density that want water-based sanitation but have no need for centralized sewer systems (Fig. 3-6 left). The septic tanks are designed for on-site treatment of domestic sewage, which is collected from

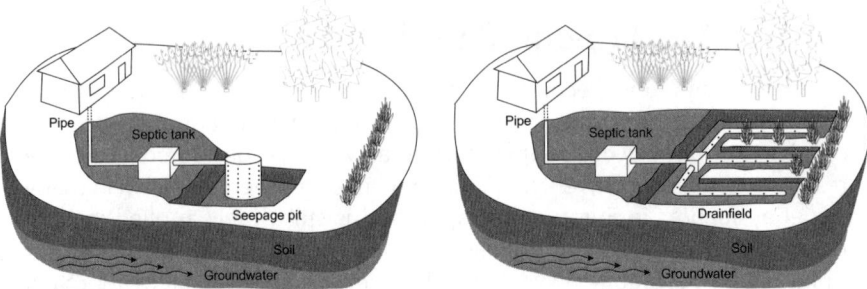

Figure 3-6. Conceptual drawing of a septic tank with a seepage pit (*left*) and with a drain field (*right*).

Source: Adapted from EPA (2002).

flush-toilet systems. The tanks are located underground and may consist of one or two compartments. Contaminants are removed from the wastewater by either settling of heavy particles or by flotation of materials less dense than water, such as oils and fats. The organic matter in the sludge and in the scum layer is digested anaerobically by bacteria. As a result, methane gas is produced, which emerges through ventilation openings in the tanks. Septic tanks can reduce the BOD of raw sewage by up to 40% and the suspended solids content by 65%; they achieve little pathogen removal but the effluent is thus much more readily absorbed into the ground than is raw sewage. Periodically, the accumulated sludge must be removed from the septic tanks. Septic tanks are easy to operate and maintain because there are no electrical requirements and no moving parts. Effluent from the septic tanks drains into seepage pits, which consist of underground pits from where the effluent percolates into the soil. A bacterial slime layer forms where effluent percolates into the soil and the microorganisms in this layer decompose some of the organic pollutants contained in the effluent.

On-Site Treatment System 5: Septic Tanks Followed by Drain Fields

Septic tanks followed by drain fields are appropriate for similar areas as for seepage pits (Fig. 3-6 right). The function of the septic tanks is similar but the seepage technique is different. A drain field is a small land area consisting of one or several long trenches into which septic tank effluent is discharged through underground perforated pipes. The sewage percolates into the ground, where bacteria in the soil decompose some of the organic matter. If constructed properly, no maintenance is required, but the trenches might clog and then require maintenance work. Drain fields provide a better disposal method than seepage pits but are not as easy or cheap to construct and require more land (UNEP 2002a). A case study of septic tanks followed by drain fields is provided in Chapter 4.

On-Site Treatment System 6: Septic Tanks Followed by Subsurface-Flow Constructed Wetland or Sand Filter

Septic tanks followed by a subsurface-flow constructed wetland or sand filter are appropriate for similar areas as for the seepage pit and drain field described above. The function of the septic tank is similar but the seepage and evaporation technique is different. This on-site system produces high-quality effluent suitable for dripline irrigation into or onto land within landscaped areas, or for providing a source of reclaimed water for recycle uses. Wetland plants are grown in aggregate, with the effluent water level maintained just below the aggregate surface. An improved large-volume grease trap should normally precede the constructed subsurface-flow wetland, and this grease trap will require regular maintenance (sand filter and constructed wetland technologies are elaborated further in Section 3.5.2).

On-Site Treatment System 7: Greywater Reclamation Units

Greywater reclamation units are used for recovery of bath and laundry water being recycled for toilet flushing (Fig. 3-7). Greywater is fed through a hold-

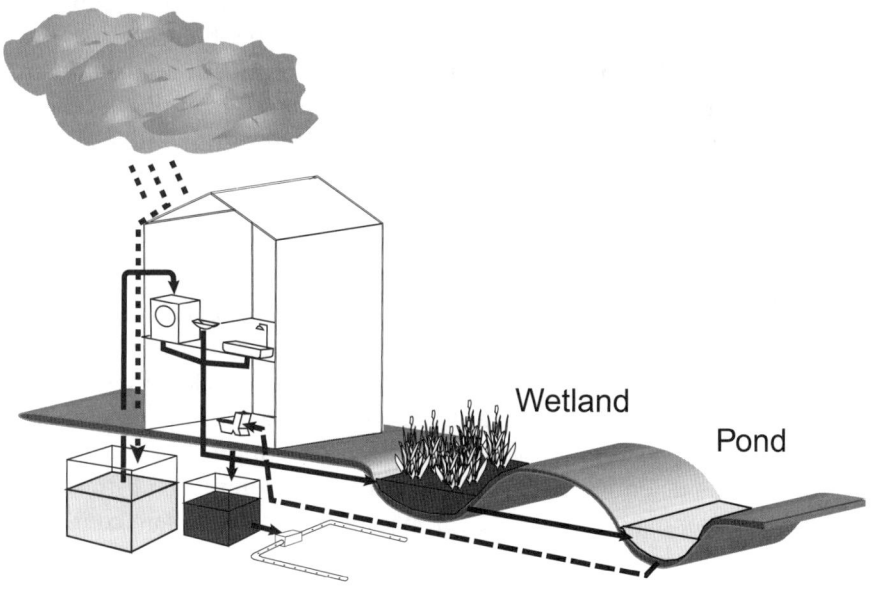

Figure 3-7. Using treated greywater to flush a toilet.

Source: Adapted from Veenstra, S. (2000). "Wastewater treatment—part 1." Unpublished lecture notes. Delft, The Netherlands, UNESCO/IHE Institute for Water Education.

ing tank (perhaps even a lined constructed wetland) before being pumped back to the toilet flushing system. The system only requires an additional small piping network and a pump.

3.5.2 Cluster and Central Treatment Systems

In cluster and centralized treatment systems, all wastewater is collected and transported to a central treatment site and then recirculated and/or reintroduced to the ecosystem. These systems tend to involve an extensive pipe network typically involving pumps and pumping stations.

For *cluster treatment*, the focus is on relatively small treatment plants designed to service a group of houses or businesses; a number of cluster treatment plants would be needed to service a whole urban area. Conversely, *centralized systems* refer to the management of wastewater in one (or a few) treatment plants servicing a whole city. Cluster treatment systems provide considerable flexibility. For example, a community or city may decide that it wants to continue with on-site treatment but, at the same time, allow developments of a certain size that cannot be serviced by on-site systems to utilize cluster systems. A cluster system may also allow a more managed land-based ecosystem re-entry because the volumes of wastewater treated will be relatively small compared to a citywide treatment system.

Primary treatment in cluster or centralized systems can be accomplished in a communal septic tank equipped with effluent outlet filters, or in a two-tiered Imhoff tank, which provides a better and more reliable effluent quality and is more economical to operate because of its capacity to hold sludge and decrease its bulk via digestion. Wastewater treatment can be provided via a range of centralized treatment options. Those that are most appropriate for developing countries are outlined below.

Cluster/Central Treatment System 1: Ponds

Ponds, also referred to as *lagoons* and *waste stabilization ponds*, are appropriate for waterborne sanitation systems in warm climates and in areas where land is available and relatively cheap, as in suburban areas. Ponds are also appropriate for treatment of sludge from on-site systems. Pond systems can accept widely varying input loadings due to the buffering action of their storage volume and detention time. Ponds are the most common full-treatment system in developing countries. A typical pond treatment system consists of three to five ponds in series, where the first pond is anaerobic, the second facultative (i.e., having a combination of both aerobic and anaerobic activity. In its top zone it is aerobic, whereas it is anaerobic at its lower zone), and the third, fourth, and fifth are maturation ponds. Some pond systems consist of several cells in parallel, with each cell having 5 to 10 days' retention capacity.

The cells-in-series configuration improves the efficiency of bacterial removal. Advantages of the pond system include low capital cost; low O&M costs; good effluent quality if designed and operated properly because they provide BOD, nutrient, and pathogen removal; and simple operation that does not require skilled operators. However, periodic removal and treatment of bottom sludge is required, typically every 10 to 20 years.

Ponds can be filled with floating macrophytes such as duckweed or water hyacinth. This plant material can be harvested and used as animal feed, thus recycling the nutrients from the wastewater. Duckweed-based wastewater treatment has been successfully introduced in a number of countries. In Bangladesh, a local NGO is operating a small-scale, duckweed-based pond for the treatment of domestic sewage. The protein-rich duckweed biomass is harvested daily and fed to adjacent fishponds, thereby combining a cost-effective treatment with revenue-generating aquaculture (UNEP 2002c). A case study of ponds is provided in Chapter 9.

Cluster/Central Treatment System 2: Trickling Filters

Trickling filters are appropriate for relatively wealthy, densely populated areas with a fairly constant population to maintain uniform loading. Advantages of trickling filters include high effluent quality in terms of BOD and suspended solids removal; low operational costs (low electricity requirements); and simpler processes compared to activated sludge or package treatment plants. Trickling filters consists of a rock or gravel medium where organisms grow in a thin biofilm. Presettled wastewater is trickled over the surface of the filters, often by use of rotating distribution pipes. Oxygen is thereby obtained by

---

))) Box 3-8. Pond Systems in Harare and Gaborone

Large-scale treatment technologies in Africa are few and seldom successful. Only 2% of cities in sub-Saharan Africa have wastewater treatment and only 30% of these are operating satisfactorily. Two exceptions are Harare, Zimbabwe, and Gaborone, Botswana. Harare is unusual in the degree to which its wastewater is treated: its five pond treatment plants provide treatment for at least half of the city's wastewater, and discharge is diverted to municipal farms for irrigation of pastures and crops. The treatment system in Gaborone is based on waste stabilization ponds, and 18,000 to 75,000 m^3 of wastewater per day is deposited in these ponds covering 52 hectares. Treatment occurs through natural processes, with no machinery or energy input except for solar energy. This has resulted in a reasonably high treatment standard. Some of the treated wastewater in Gaborone is also re-used for irrigation (UNEP 2004).

direct diffusion from air into the biofilm. *Biofilters* and *rotating biological contractors* (RBCs) are systems that build on similar processes as trickling filters. Skilled labor is required to keep trickling filters operating trouble-free (e.g., to prevent clogging), ensure adequate flushing, and control filter flies. RBCs are discussed in further detail in Chapter 7, Section 7.3.

Cluster/Central Treatment System 3: Sand Filters

Sand filters, or *depth infiltration systems*, may be appropriate for relatively wealthy, densely populated areas, hotels, and tourist resorts, especially if there is an opportunity to re-use treated effluent. These systems cope well with fluctuating loading rates and produce a high effluent quality because they reduce bacteria numbers and significantly reduce organic matter and suspended solids. Sand filters are relatively economical to construct because of their reduced size, but pumping costs for dose loading are higher, and regular backwashing to prevent clogging of the filter medium is required. The most common filtering medium is sand, but anthracite, synthetic fiber, and crushed glass are also used.

Cluster/Central Treatment System 4: Constructed Wetlands

Constructed wetland systems are regarded by some as an extremely promising wastewater treatment technology for developing countries (Nelson 2002). There has been increasing interest in using constructed wetlands for wastewater treatment since early studies demonstrated their effectiveness at removal of nutrients and suspended solids. Also, constructed wetlands show increased rates of uptake in warmer climates and such systems operate even more efficiently in most developing countries. The effectiveness of subsurface-flow, gravel-bed wetlands, especially the vertical-flow systems, has been substantially improved in recent years. There are three types of constructed wetlands: horizontal subsurface-flow, vertical subsurface-flow, and surface-flow (Fig. 3-8).

Subsurface-flow systems have demonstrated their appropriateness in situations of small on-site or clustered wastewater loadings; in areas where land is scarce (subsurface systems require only one-fifth the area compared to a surface-flow wetlands); in situations where avoidance of malodor and mosquito-breeding are important; in coastal areas with groundwater too close to the surface, such as often occurs during the wet season; and in sites with rocky or impermeable clay soils that prevent standard leach fields from operating. Subsurface flow involves effluent treatment via flow through a porous medium: one type has a horizontal flow of wastewater from one end of the medium to the other, and the other has wastewater being pulse-pumped onto the full top surface of the medium to then flow vertically down toward the outlet.

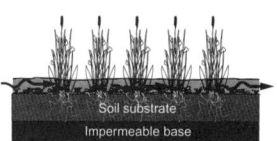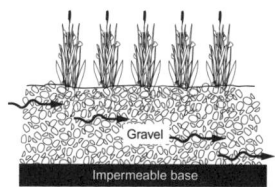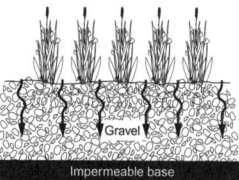

Figure 3-8. Conceptual drawing of (*left to right*) a surface-flow constructed wetland, a horizontal subsurface-flow constructed wetland, and a vertical-flow constructed wetland.

The advantages of the subsurface-flow wetland approach include 99% reduction of fecal coliform bacteria without the use of expensive, environmentally harmful chemicals like chlorine; 85% to 90% BOD reduction and substantial removal of nitrogen and phosphorus; the systems are low-cost, low-tech, and long-lived; and they have simple O&M requirements. Also, this approach has the potential to reach higher treatment levels by increasing the wetland area, providing the equivalent of advanced water treatment; significantly less wastewater can be discharged (35% to 70%, depending on design) because plants use large quantities of water in their transpiration; and landscape can be beautified, such as with botanical garden displays or creation of wetland ecosystems with rich biodiversity, wildlife and bird habitat, and growth of plants for use or sale (Nelson 2002). Case studies of subsurface flow-constructed wetlands are provided in Chapters 5, 6, 8, and 10.

))) Box 3-9. Constructed Wetlands in Egypt

One of Egypt's most pressing environmental problems is the lack of clean, reliable freshwater. Much of the heavily polluted water flowing through the Nile River enters large coastal lakes, such as Lake Manzala, before pouring into the Mediterranean Sea. Wastewater has traditionally been left untreated, degrading the lake and its once prolific fisheries, and sending pollution downstream into the Mediterranean coastal ecosystems. The government has initiated a project to treat and re-use wastewater for productive purposes through the use of constructed wetlands. The project involves a local community in the maintenance of the facility. Every day, 25,000 m^3 of polluted river water is pumped from the Bahr El Baqr canal into a series of large ponds, where toxic sediments settle out. The water then flows into constructed wetlands where it is filtered by plants and bacteria which gradually remove additional pollutants. The entire process is chemical-free and costs one-tenth of conventional technologies (UNDP 2005).

Surface-flow wetlands have features similar to the subsurface-flow types described above, but require larger land area to provide the same treatment efficiency and do not control odor issues as well. Surface-flow wetlands provide efficient treatment over a 5- to 10-day flow-through (retention) period. Water treatment occurs through settling and bacterial growth on the stems of emergent wetland plants that are rooted in the soil on the bottom of the shallow pond, as well as by aeration of the water by oxygen transfer processes.

Cluster/Central Treatment System 5: Overland Flow
Overland-flow treatment systems offer both a treatment function and an ecosystem re-entry method. These systems are appropriate in urban or suburban areas that are close to commercial rural or forest areas with cheap and readily available land. Treatment in overland-flow systems occurs within the topsoil mantle. To ensure that the aerobic renovation capacity of the soil is maintained, alternating cycles of load and rest are required. Effluent to be treated is spread over the upper surface of a sloping, grassed plot and is treated via sheet flow as it moves down to a collection system at the lower edge of the plot. As the wastewater flows over the land, some will be infiltrated into the soil, achieving re-entry to the ecosystem. Flow that does not soak in is collected as polished effluent for disposal in nearby waterways (Fig. 3-9).

Overland-flow systems work by soil and plants acting as filters that trap and treat, through various mechanisms, contaminants in the wastewater and allow the remaining wastewater to drain through the soil profile. The

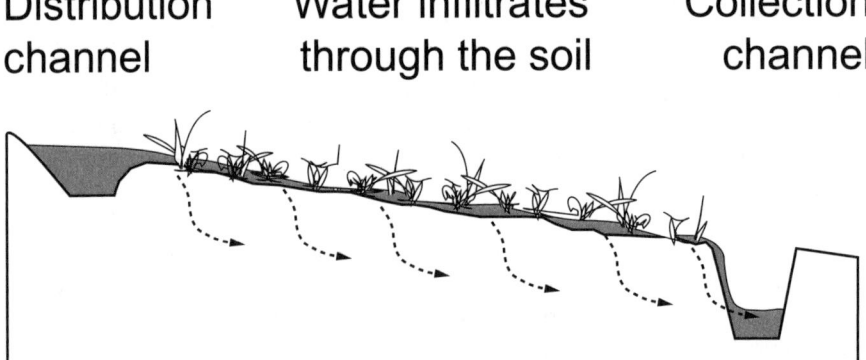

Figure 3-9. Conceptual drawing of an overland-flow treatment system.
Source: Adapted from UNEP (2002c).

net effect is a beneficial system allowing for both the effective remediation of wastewater and the recycling of water, nutrients, and carbon via biomass production.

3.5.3 Combinations of On-Site, Cluster, and Centralized Treatment

On-site systems such as septic tanks followed by seepage pits are often seen as old-fashioned systems that should be replaced when possible, whereas fully centralized systems are seen as modern, desirable systems. This perception is changing, however, as the possibilities of cluster systems become better known and acknowledged.

On-site systems can seem a bother for landowners because they require considerable direct care; furthermore, they are coming under increased scrutiny, especially by agencies concerned about public health, because in developing countries such systems are often poorly maintained and operated, thus creating problems with discharges to land and waterways, and contamination of the water supply. Sometimes these failures are caused by lack of information about how to operate and maintain the systems; other times it is a matter of cost and poverty. On-site system failures can push a community into choosing off-site cluster or centralized systems despite the fact that the local soils can still handle on-site systems—changing systems would not be necessary if the on-site systems had just been better managed. Accordingly, there have been recent attempts to place on-site wastewater systems under integrated management programs, particularly with respect to their operation, maintenance, and monitoring.

It is important to note that these three categories of treatment systems do not necessarily represent moving from the less sophisticated to the modern. Each one is equally important and capable of delivering safe, efficient water treatment. The real issue is which system best fits the specific local social and historical context and its environment, especially soil conditions, water quality, and ecosystem sensitivity.

Most cities in developing countries have on-site systems, and the decision making should center around how these work, whether they can work better, and what the options are for combinations with cluster or centralized systems. For this reason, a strong focus must be maintained on on-site system configurations in relation to required standards for effluent discharge. Case studies of combined systems are provided in Chapters 5, 6, 7, 8, 9, and 10. Table 3-1 provides a summary of some of the effluent qualities reached by various on-site, cluster, and centralized treatment technologies.

> **Box 3-10. A Combined Wastewater Management System**
>
> Forty new residential lots in Golden Valley, New Zealand were designed and constructed as a combined on-site and cluster system. The system included a pumped modified effluent drainage servicing (MEDS) collection system, where filtered septic tank effluent is conveyed in 50-mm pressure sewer lines from a pump within each septic tank to a central recirculating sand-filter treatment plant located in an enlarged and landscaped central median strip on the access road serving the development. Some of the high-quality effluent produced is disinfected and returned to each lot as nonpotable reclaimed water for toilet flushing; some nondisinfected effluent is pumped to an area of steep terrain that is irrigated by driplines into eucalyptus-planted plots; and the remainder is held in storage for fire-fighting purposes. The advantage of the sand-filter treatment system is that it can run on a modular basis. Treatment capacity can be extended to match housing numbers as constructed over time. On a seasonal basis, modules can be started up and shut down to fit the expansion and contraction of holiday occupancy. All this can be accommodated while maintaining consistently high treatment performance (ME/NZ 2003).

3.6 Element 3: Energy Consumption

High energy consumption is often the Achilles heel in the operation of wastewater management systems in developing countries. The systems may lack overall justification, may be overdimensioned, or may utilize too advanced technology—whatever the cause, the first thing to happen is a wish to save O&M costs, and the first in line is the electricity costs. Pumps are turned

Table 3-1. Performance of Different Treatment Technologies

	Raw Domestic Wastewater	Septic Tank	Sand Filter	Constructed Wetlands
BOD, g/m^3	200–300	120–150	5–15	5–15
Suspended solids, g/m^3	260–400	40–120	5–20	5–20
Total nitrogen, g/m^3	30–80	40–60	30–50	5–30
Total phosphorus, g/m^3	10–20	10–15	5–10	5–10
Fecal coliform, cfu/100 mL	10^6–10^8	10^3–10^5	10–10^3	300–1,000

Source: Modified from UNEP (2002c).

off, aerators are stopped, treatment is by-passed, and the electricity bills get smaller. Appropriate and sustainable wastewater management systems must maintain a strong focus on having the lowest possible electric costs.

The energy issue can also be seen from a global "recovery of energy" perspective. Appropriate and sustainable wastewater management systems should consider the energy component of wastewater and that of wastewater treatment systems. Conventional wastewater treatment, such as activated sludge, requires substantial inputs of external energy, usually coming from nonrenewable sources. Theoretically, 0.8 m^3 and 3.0 m^3 of oxygen are required for the oxidation of 1 kg of organic matter and ammonia, respectively. In aerated systems, several times this volume must be forced into the water phase at the expense of valuable energy. However, the treatment of wastewater in a high-rate anaerobic reactor does not require oxygen input and, in addition, will yield some 375 L of methane per kilogram of BOD digested. About 90% of the energy contained in organic matter will end up as methane gas. This is not only positive for the overall energy balance of the system, but also replaces an equivalent amount of nonrenewable energy and greenhouse gas emissions if the methane is used as an energy source (UNEP 2002c).

The optimal and most appropriate and sustainable systems are fully powered by gravity, solar, wind, biomass, waves, or other renewable energy sources. Here are three of many possible examples of sustainable energy options.

Energy System 1: Gravity-Based Systems
On-site systems are often fully powered by gravity. Water flows from the house to the treatment system and then on to the seepage, drain, or constructed wetland system. For larger cluster or centralized systems, gravity flow can also be achieved, especially in steeper topographical areas or where it has been carefully designed into the systems. Case studies of full or partial gravity-based systems are provided in Chapters 4, 5, 8, 9, and 10.

Energy System 2: Solar-Powered Pumps
Improvements in solar power systems continuously lower their costs, increase their efficiency, and extend the life span of both solar panels and batteries. Solar-powered systems have become increasingly easy to operate (e.g., solar pump package systems). In addition, innovation opens up new ways of integrating solar panels in architecture and the urban environment. The current (somewhat rigid) rectangular box design is being superseded by photovoltaic cells integrated into building façades, window panels, or roofing materials. However, doubts can be raised about many present-day solar-powered pump systems because they are often too complex and too expensive, thereby reducing their sustainability. A traditional solar-powered

pump system consists of solar modules, a charger controller, a battery, a backup generator, and a bidirectional inverter/charger. If investment costs are included, the power they supply is often much more expensive than traditional power supplied by a grid. Running solar-powered pumps is typically only cheaper than the grid if the units are donated. A case study of solar-powered pumps is provided in Chapter 6.

Energy System 3: Siphons
A siphon-based distribution system is a simple but practical technology that reduces overall system complexity as well as energy consumption. Large siphons can be used instead of electrical pumps for the required intermittent pumping of wastewater to clustered vertical-flow constructed wetlands or to seepage pits from on-site septic tanks. A siphon is a simple mechanism that triggers the release of water when a certain water level difference is reached between the intake and the outlet of the siphon. Once the water at the siphon intake has reached the trigger level, full flow is activated and water is discharged until the water level in the reservoir reaches a low level and the siphon starts taking in air, which stops the flush action. Through appropriate design of the siphon, the flow rate can be determined with fairly good accuracy. A siphon contains no moving or mechanical components and is a robust and reliable mechanism. For a constructed wetland technology, the development of cost-effective and reliable methods for intermittent pumping is very important because vertical-flow systems are much more effective than the other types of constructed wetland systems. A case study of siphons for constructed wetlands is provided in Chapter 6.

⟫⟫ 3.7 Element 4: Urban Integration

If carefully designed, the wastewater management system can be fully integrated into the urban environment and become a part of the city, the urban landscape, and the community. Traditionally, wastewater management infrastructure is ugly, heavy, and smelly, and therefore must be hidden as far away as possible. However, this does not have to be the case.

The possibilities for urban integration are numerous. Integrating wastewater management systems into the layout of housing estates creates an opportunity for a lush, green environment. The systems can be constructed to utilize stormwater, greywater, and even black wastewater through various elements such as seepage systems, ponds, constructed wetlands, and subsurface irrigation systems for open green areas.

The key is to combine quantitative parameters such as area demand, performance, leveling, and land availability with qualitative parameters such as aesthetics, social integration, and usability. The treatment system can be designed to address both odor and beautification. In the future, we hopefully will see more examples that combine these quantitative and qualitative parameters—making wastewater facilities both visible and invisible, and making the visible more pleasing while still being effective, integrated, and safe. Integrated water and wastewater management should therefore be considered at the earliest stages in the planning of new housing estates and urban areas. Case studies of the five following urban integration techniques are provided in Chapters 5, 6, 8, and 10.

Urban Integration 1: Integration by Invisibility

To address the odor issue, the collection and treatment system must be designed to be as effective and *imperceptible* as possible. Examples of these techniques include closed-loop, separate collection pipes from each house to the treatment plant; small, underground, odorless pumping stations; and using subsurface-flow constructed wetlands as the main treatment technology.

Urban Integration 2: Multifunctional Integration

Multifunctionality of the treatment or re-use system can optimize land use and facilitate spin-offs with mutual benefits for the municipality, landowners, residents, and tourists. This type of system can collect, treat, and re-use the wastewater, thus ensuring public health and a clean environment, It can also function as, for example, a public park with walking paths, benches, and pavilions. The treatment location could include a volleyball field or could be integrated with public parking spaces—the possibilities for multifunctionality are numerous.

The use of integrated duckweed-based treatment systems illustrates the ample possibilities for urban multifunctional integration. Anaerobic technology is used to reduce the bulk of organic and suspended matter, and the energy produced (methane gas) in the biogas digesters can be used by the community. The effluent of the anaerobic reactors can be channeled to duckweed pond facilities; the duckweed can be harvested at regular intervals and used to feed fish in adjacent ponds; and the effluent can be made available for irrigation. With the income from the products generated (energy, fish food, irrigation water), the integrated system has the potential to become a commercial enterprise generating substantial revenues.

Urban Integration 3: Symbolic Integration

Chapter 6 describes a treatment facility that was designed to symbolize a local feature—a butterfly sitting on a flower—that symbolically references the butterfly-shaped contour of its island site. The relationship between the flower and the butterfly also symbolizes a new beginning—the growth and bloom of the flower, the community, and the island.

Urban Integration 4: Aesthetics Integration
An aesthetically pleasing layout of a facility may be a key to gain public acceptance. The importance of landscape design, the aesthetic composition of plants and perspective, the choice of materials such as the pavement on the walking paths and the inclusion and design of, for example, a pavilion, the design of the entrance, the lighting, the trees, and the general appearance of the facility—all contribute to an aesthetic integration into the urban landscape and to public acceptance of the treatment system.

Urban Integration 5: Topographic Integration
The actual location of the treatment facility provides opportunities and constraints. Land might be ample or scarce, the area flat or hilly, or located near housing estates, inside a recreational area, or near the coast. The location sets the options for topographic integration. For example, to counteract a local dense urban design, a clearly defined void can be incorporated into the facility design to balance the numerous compounds in the vicinity. Open green and blooming areas can contrast with the surrounding cement structures. The landscape design can provide a focal point, a *point de vue*, for the houses and hotels along surrounding hillsides.

))) 3.8 Element 5: Re-Use and Re-Entry of Wastewater

Domestic wastewater contains essential resources such as water, nutrients, and organic material. Treated wastewater produces liquid wastewater and sludge. Both of these wastes can be processed to recover re-usable water and composted biosolids that can used for, as an example, horticultural application as a soil conditioner, and thereby become a sustainable part of a local ecosystem.

The two key issues regarding this fifth element in wastewater management system design are, first, how to make use of the treated wastewater, and second, how to secure its best re-entry into the ecosystem.

3.8.1 Re-use of Wastewater

Past and Present Re-Use of Wastewater
In rural and suburban areas of most developing countries, use of wastewater for irrigation is not a matter of choice. Canals or rivers used for agricultural irrigation carry domestic wastewater from upstream towns, and in semi-arid areas the use of wastewater-filled drains may be the only water source that supports the livelihoods of millions of poor people by irrigating high-value crops. The obvious solution—building wastewater treatment facilities—is prohibitively expensive and not an option for the poorer areas in Africa, Asia, and South America. Neither is legislation to ban the use of wastewater and sewage for crop irrigation.

The application of wastewater to land for treatment and disposal was one of the earliest wastewater treatment technologies. Livestock manure has been used as fertilizer for approximately 5,000 years, whereas humanure has been known as fertilizer since the ancient Chinese Shang Dynasty dating back 3,500 years. Land application systems have included application to edible and nonedible crops, to rangelands, to forests and wood plantations, and, more recently, to recreational areas including parks and golf courses, and to disturbed lands such as mine spoil sites.

In many areas of the world, wastewater re-use has been practiced using a combination of treatment technologies that achieve a very high degree of treatment. Many states in the American West have in recent decades been treating wastewater to tertiary treatment standards and then allowing the wastewater to be re-used for irrigation or to recharge groundwater aquifers. Although this is an effective method of treatment and re-use, it is very expensive and can rarely be practiced in developing countries.

Land application systems that utilize the land as a treatment unit and not just as a disposal area are gaining acceptance in many arid regions. These systems are cheaper to construct and operate and can be operated by personnel familiar with common irrigation systems. Many arid regions, such as Egypt and Mexico, lack infrastructure support and cannot afford expensive treatment technologies. For these regions, slow-rate land application systems may be appropriate and low-cost because properly designed land application units provide environmentally safe wastewater disposal by removing pathogens, nutrients, and suspended solids. Also, the wastewater can be used to create value-added benefits such as wetlands, crops, trees for fuel wood, pulp products, lumber, cotton, and restoration of dryland desert ecosystems.

Increased Re-Use of Wastewater in the Future
It can be expected that re-use of wastewater will increase in the future due to the preference for wastewater re-use over effluent discharge; the increasing pressure on existing water resources due to population growth; increased agricultural demand; climate change (increasing temperatures); the growing number of successful wastewater recycling projects; and the increased costs associated with operating wastewater treatment plants to meet higher quality standards.

However, re-use of wastewater and sludge is a relatively new dimension of wastewater management systems, and concerns have been raised. Health authorities have concerns regarding re-use of wastewater because of the possibility of direct contact with pathogens if something goes wrong with the treatment process or if the system is not adequately maintained. Concerns have also been raised about wastewater irrigation being applied directly to food crops, and there is uncertainty about compost as an end use. However, it

> Box 3-11. Re-Use through Land Application in Africa and Latin America

Wastewater re-use in Africa is still in its infancy, but the last decade has seen an increased number of re-use projects. Agriculture is usually the principal water user, followed by industrial, retail/service, and domestic use. In several semi-arid areas of Africa, water allocation is critical and recycling of wastewater is becoming a high priority. In these dry zones, wastewater may constitute 25% to 75% of the available irrigation water. Examples include Angola, Namibia, Botswana, Zimbabwe, South Africa (with 16% or 70 million m^3/yr of wastewater re-used) and Tunisia (with 75% or 68 million m^3/yr of wastewater re-used). In Egypt the Ismailia Serrabium wastewater treatment plant, built by the government of Egypt and the United States Agency for International Development (USAID) in 1995, is re-using its treated effluent. Ismailia has a population of about 500,000 and the treatment plant receives about 85,000 m^3 of wastewater per day. The adjoining land application facility uses a land area of about 200 hectares with up to 2,000 more hectares available, and supports nursery and grow-out operations where all plants are drip-irrigated with the treated wastewater. The nursery production capacity at Serrabium Forest is 100,000 trees per year.

In Latin America, with important exceptions in Peru, Argentina, Chile, and Mexico, wastewater re-use is not widely applied. Wastewater re-use in Latin American countries is mainly confined to agricultural lands, where it is estimated that about 500,000 hectares have been irrigated—some 20% to 25% of the overall worldwide figure. This includes Santiago in Chile, with about 16,000 hectares wastewater-irrigated; 4,000 hectares in Argentina; and large re-use schemes in Mexico and Peru. In the outskirts of Mexico City, lands irrigated by wastewater re-use comprise nearly 90,000 hectares, together with another 275,000 hectares spread throughout the country. In the Eastern Mexico Valley Basin, 14,500 hectares planted with 50,000 trees have been wastewater-irrigated since 1971. Beginning late in the last century, more than 70,000 hectares of land in the Mezquital Valley have been wastewater-irrigated. This represents the world's largest area of wastewater-irrigated agriculture: the land is assigned to 45,000 families who grow corn, oats, beans, wheat, pumpkins, tomatoes, and so forth. The wastewater storage capacity of 350 million m^3 is divided among six reservoirs, which convey treated water into more than 1,800 km of channels and canals. About 45 m^3/s of wastewater is used for irrigation; it is estimated that the economic value of wastewater-irrigated crops is about $100 million USD yearly, and the environmental benefits include a reduction of 1,150 tons per day of BOD load.

On the desert coast of Peru, wastewater irrigation projects have been developed for approximately 10,000 hectares of land in the Lima area, which has about 7.5 million people. However, the raw wastewater used for

> irrigation in this region is often used to irrigate crops for human consumption, especially fresh produce such as salad crops and vegetables, and has caused serious public health problems, including diarrhea, intestinal fevers, hepatitis, and parasitosis, mostly among poor citizens. In Lima, diarrhea is the most common cause of infant mortality and its typhoid fever rate is the highest in Latin America (UNEP 2002c).

seems to be commonly agreed that nonpotable water use is acceptable if it is not used for food production and if it first passes through soils.

Another serious problem relates to the mixing of domestic and industrial wastewater. The return of nutrients from the urban areas to the soils of the agricultural areas from whence the foods derive is how the nutrient cycle would be most completely closed. However, one of the current practices that make re-use in urban environments difficult is the indiscriminate mixing of domestic with commercial and industrial wastewater. Some of the common hazards of wastewater re-use relate not to the common constituents of domestic wastewater, but to toxic constituents that are primarily found in industrial and manufacturing wastewater. These substances include heavy metals and complex synthetic or organic compounds (e.g., organochlorides and dioxins). These substances are both expensive to test for and pose hazards if disposed onto land or into groundwater and coastal ecosystems.

These considerations suggest that a core principle in appropriate and sustainable wastewater management is to maintain separate collection systems, where industrial and other wastewater are not mixed, so that concerns about such toxic compounds can be avoided and re-use can be implemented more effectively. They also suggest that governmental support and control are required to create an effective and safe environment for the re-use of wastewater.

Until now, wastewater has mainly been re-used in the following five key sectors. It can only be expected that re-use of wastewater will continue to increase in these sectors.

Re-Use in Agriculture

Agriculture consumes large quantities of water, and recycled wastewater has been used in a variety of applications, including crops such as fruit, vegetables, cotton, and sugarcane; pasture production and turf farms; horticulture such as plant nurseries, vineyards, and cut flowers; and forestry. Forest and grassland effluent irrigation systems commonly utilize effluent spray irrigation management with the advantage that nutrients and water enhance tree growth. Where dripline systems are used, buffer distances can be very small and horticultural use of the treated effluent nutrients and water becomes feasible. In

> **Box 3-12. Government Support of Re-Use Schemes**
>
> The importance of and tools for governmental support of wastewater re-use can be illustrated by the government of the state of Queensland (Australia), which provides policy, financial, and technical support for such projects. Its *policy* states that "[W]here it is safe, environmentally sustainable and cost-effective, the Queensland Government strongly encourages the recycling of treated effluent in preference to discharge to waterways, estuaries or ocean," and that it supports "on-site greywater recycling in unsewered areas; on-site blackwater recycling in unsewered areas; regulated trials of on-site greywater recycling in sewered areas; urban stormwater recycling; and rainwater tanks." *Financial support* is provided through several mechanisms: a subsidy of up to 50% for planning, design, and construction of re-use projects that obtain treated effluent from wastewater treatment plants as an alternative to discharging that effluent to coastal waterways; a smaller communities assistance program, focusing on communities with less than 5,000 people, provides a subsidy of up to 100% of the costs of water supply and sewer services, and includes provisions for wastewater recycling; and an advanced wastewater treatment technologies program to encourage the introduction of new and/or innovative wastewater treatment technologies. Finally, the Queensland government developed a detailed action plan that encompasses, among others, changes in existing state *laws* to support wastewater re-use, provision of *guidelines* for wastewater re-use, and re-use *demonstration projects* to raise community awareness (EPA/Q 2001).

terms of volume, the greatest potential lies in irrigation of pasture, field crops, and tree crops.

Re-Use in Industry and Business

Industry and businesses have re-used wastewater for a number of purposes. In industry, it has been used for cooling in a variety of processes: for boiler-feed water, process water, wash-down and cleaning, flushing toilets and urinals, dust suppression, and irrigation of grounds. In businesses, wastewater has been re-used in commercial car washes, paper mills, mines, petroleum refineries, power stations, manufacturing of concrete, bricks, textiles, metals, and paint, road construction, tanners and hide curing, tourist resorts, and distilleries and wineries.

Re-Use in Houses

For individual and clusters of houses, recycled wastewater has been used for toilet flushing, car washing, cleaning, and garden irrigation. It can be expected that dual reticulation might increasingly be applied for the re-use of wastewater. This involves the supply of water from two separate sources, using two sets of pipes. One set provides clean water for drinking, cooking, bathing, and laundry; the other provides recycled wastewater for other purposes such as garden irrigation.

Re-Use in Recreational and Open Spaces
Examples of recreational and green-space usage of treated wastewater include the irrigation of open spaces such as golf courses, sports fields, resorts, cemeteries, parks, freeway landscaping, urban beautification, new water features, and for a variety of recreational purposes such as artificial lakes for boating. For example, about 75 golf courses in Queensland, Australia are irrigated with recycled wastewater, accounting for 45% of the water that is recycled from municipal wastewater treatment plants in the state (EPA/Q 2001).

Re-Use for Environmental Purposes
Environmental use of recycled wastewater includes restoring riverine environments such as wetlands that have been degraded as a result of altered or reduced streamflows; constructing new wetlands; and creating ornamental lakes designed for wildlife habitats or for aesthetic purposes. The use of treated wastewater for wetland restoration has been especially advantageous because it involves artificially recharging water back into a wetland to offset the loss of water from drainage of surrounding areas and the lowering of the water table.

3.8.2 Re-Entry of Wastewater into the Ecosystem

Not long ago, the way wastewater re-entered the environment was not a major focus for wastewater management planners. For on-site systems the main concern was to ensure that septic tank fields were able to absorb the wastewater; periodically, the tank would need to be cleaned out and the sludges buried. Various levels of treated wastewater from centralized systems would be discharged directly into rivers, coastal waters, or the sea. Untreated wastewater would often be discharged via sewer outfalls onto coastal areas. However, in the last decade focus has shifted markedly from water-based to land-based re-entry systems.

Land-Based Re-Entry
Land-based re-entry means that treated wastewater is returned to land by various irrigation methods, such as seepage into the soil subsurface, flood irrigation, overhead sprinklers, subsurface drippers, and evapotranspiration. Options for returning treated wastewater to the ecosystem within the site boundaries depend on wastewater quantity and quality, and on-site conditions such as soil types, area, and slope of land available, location of groundwater, and local

> **Box 3-13. From Water-Based to Land-Based Re-Entry**
>
> A survey of the main forms of community wastewater effluent re-entry in 283 New Zealand communities showed that 73% had re-entry into waterways, whereas 27% utilized land-based re-entry systems. The survey concluded that water-based re-entry systems often do not provide sound environmental performance; this has shifted the emphasis for new or upgraded facilities away from water-based re-entry toward land-based re-entry. This shift in approach has been particularly significant for smaller communities because the land areas required could readily be found in adjacent rural areas. For large communities, the strategy for upgrading their treatment and ecosystem re-entry systems involves the use of constructed or natural wetlands as an appropriate buffer between the treatment plant and the natural water into which the final discharge diffuses (ME/NZ 2003).

climate. It also depends on the sensitivity of the waterways and coastal ecosystems and the relative importance of these ecosystems' goods and services.

Wastewater land-based re-entry technologies can roughly be divided into surface and subsurface seepage systems. Various technologies include:

- *Subsurface seepage systems.* A large number of different subsurface seepage systems exist. *Subsurface seepage systems (seepage pits* and *drain fields)* are commonly installed in developing countries because they are simple, cheap systems and only require subsoils with appropriate drainage characteristics and not-too-high groundwater levels (see also Section 3.5.1 on on-site treatment technologies). *Low-pressure effluent distribution trenches* are specially designed shallow and narrow trench systems with a nested perforated dosing pipe within a drain-coil line. They are used for either deep, sandy soils to distribute septic tank effluent for further in-soil treatment, or for deep topsoil conditions overlying clay to distribute effluent for topsoil treatment and evapotranspiration. *Evapotranspiration seepage beds* are appropriate where soils have impeded drainage, and are used in climates with good evapotranspiration rates and lower rainfalls. Beds and/or surrounding spaces between beds are planted with high-transpiration shrubs, plants, and/or grasses. *Subsurface dripline irrigation* utilizes driplines laid within topsoil to depths of 50 to 100 mm. A recent technological improvement is the controlled-drip subsurface dripline system that provides a geotextile wick above a plastic strip to ensure that effluent disperses fully along the length of the dripline instead of concentrating at the drip emitters.

- *Surface seepage systems.* Many different surface seepage systems exist. *Surface spray irrigation* is typically used for wastewater that has received secondary treatment (e.g., from constructed wetlands) and disinfection via ultraviolet light or chlorine tablets. *Surface dripline irrigation* includes driplines laid on the soil surface and covered with mulch, bark, or compost. These systems can be designed for incorporation within a landscaped area on the lot. Septic effluent drip-irrigation is being trialed in many projects with the objective to provide more effective distribution of primary effluent into aerobic topsoil layers to take advantage of the soil's treatment capacity. *Rapid infiltration systems* function both as treatment and disposal because partially or fully treated effluent is soaked into the ground at a high rate for further in-soil treatment. Only sandy soils are suitable for long-term use and the water table must be sufficiently deep so that all pathogens are trapped in the soil, where they can gradually die off and not contaminate the groundwater.
- *Land-based sludge disposal.* Land based re-entry systems also include disposal of *sludge* to a landfill site, spread onto land, composted, pelletized, or treated for use as a soil conditioner. There is a growing focus on converting sludge into biosolids and reducing the level of water in them to lessen handling problems when they are disposed to landfills or used as soil conditioners. The wet biosolids may be dried on special sand beds before being collected as dried cake for trucking to a landfill or, alternatively, may be spread on land for agricultural or forestry fertilization.

In on-site systems, accumulated sludge should periodically (every 6 months to several years) be removed from the system by vacuum trucks and transported to a central sludge management facility. Unfortunately, in many developing countries the vacuum truck drivers dispose of the sludge at the nearest convenient location (UNEP 2002c), which may well be a watercourse, a riverbank, or a coast.

Water-Based Re-Entry

Water-based re-entry is the still most commonly used method of re-entry of untreated and treated wastewater into the ecosystem, and implies that wastewater is returned to the ecosystem through direct-point discharge to waterways and coastal ecosystems such as wetlands, estuaries, or the sea. Direct discharge of untreated wastewater to waterways most often results in uncontrollable impacts on the ecosystems. If wastewater is being directly discharged to waterways, high discharge standards are required and efficient wastewater treatment facilities must therefore be installed and operating. Water-based re-entry systems, including deep-sea outfalls, are acceptable at present but such

> **))) Box 3-14. Untreated Wastewater Discharged Directly to Coastal Ecosystems**
>
> Only 5% of the inhabitants of metropolitan Lagos, Nigeria, are connected to waterborne sewer systems and associated wastewater treatment plants. Existing plants do not treat the wastewater to acceptable standards and they are poorly maintained and operated. Open stormwater drains are common and in many cases act as open sewers, particularly for the conveyance of greywater. Most industrial wastewater is also discharged directly into waterways without any form of treatment. Major drains are not maintained, and both secondary and tertiary drains are poorly maintained and hence fail to alleviate flooding. In other words, almost all wastewater is discharged directly and untreated into the coastal waters or the Lagos Lagoon. The volume of wastewater generated is expected to reach 1.7 million m^3 per day by 2010 (UNEP 2002c).

systems are not feasible in the long term. We believe that future developments within the wastewater management sector in general should not focus on the implementation of water-based re-entry systems.

))) 3.9 Element 6: Organization and Finance

Appropriate and sustainable organization and financing of wastewater management systems may be seen from two perspectives: the individual or local perspective of the wastewater management system, and the national or central perspective.

3.9.1 Local Perspective

From the local perspective, or the perspective of the individual wastewater management system whether on-site, clustered, or centralized, three key issues are of importance. For a specific wastewater management system to be organizationally and financially sustainable, (1) the investment and recurrent O&M cost must be kept down; (2) sufficient and continuous income (cost recovery) must be generated; and (3) sufficient local organizational and human resources must be available. Each of these requirements for sustainability and appropriateness is discussed below.

Local Requirement 1: Low Investment and Operating and Maintenance Costs

Developing countries and international financing institutions are beginning to recognize that poor urban residents cannot afford, or necessarily want or need, costly conventional wastewater management systems. Beyond the dense urban centers, the average cost of a conventional residential sewer system may range from $500 USD to $1,000 USD and conventional treatment processes may cost $0.25 USD to $0.50 USD per /m^3. This is clearly too expensive, as many households in developing countries have annual incomes below $500 USD.

However, a broad range of cost-effective technological options are available to respond to the demands of suburban and rural areas beyond the urban center, with the potential to reduce investment costs to $100 USD or less per household, and recurrent treatment costs by at least one-half (UNEP 2002c). Low-cost treatment approaches range from on-site systems such as combined septic tank and land application systems, to cluster systems such as ponds and constructed wetlands systems. For example, the capital cost of a recently constructed wetland in Mexico was one-third the per-capita cost of a conventional wastewater treatment plant, and O&M costs were nine times lower (Nelson and Tredwell 2002). Table 3-2 illustrates the relative difference in capital and recurrent cost requirements for different wastewater technologies. Land-based options for suburban and rural areas require consideration of the availability and cost of the land, and economy of scale.

Table 3-2. Relative Costs for Different Wastewater Management Technologies

	Capital Cost Ratios	Annual Recurrent Cost (% of Capital Cost)
Pit latrine	0.28	5.1
VIP latrine	0.55	2.6
Pour-flush latrine	0.53	2.7
Septic tank	1.00	8.9
Seepage pit	1.70	0.0
Drain field	2.50	0.0
Conventional sewer	5.29	4.6
Simplified sewer	2.27	16.0
Communal septic tank	0.20	10.0
Primary treatment	0.45	9.3
Waste stabilization ponds	0.74	2.5
Activated sludge treatment	1.80	6.8
Sludge treatment	0.52	25.0

Source: Adapted from Loetscher (1999) and UNEP (2002c).

Sufficient land for on-site systems is required to enable treatment and re-use of treated wastewater on-site, and for off-site systems for treatment facilities and re-use through nearby agriculture, horticulture, forestry, or industrial activities that present opportunities for re-use. Two points should be made. First, land is often more available than centralized planners might think, and second, new types of land-based systems require much less land than previous systems.

The *cost of land*, naturally, is an important factor in implementing cost-competitive land-based systems. In the case of high land costs, however, salvage costs should also be taken into account, making it possible for the municipality or community at a later stage to sell or convert the land to other uses. If land is available at a cost that now or in the future makes such other uses feasible, a lower-cost technology utilizing more land should be chosen rather than a higher-cost technology using less land area.

Economy of scale should also be considered. Individual on-site systems do not normally present an opportunity for economy of scale, contrary to off-site wastewater management systems where the cost of treatment per unit volume of wastewater will decrease with an increase in population served (even though the cost of collection will increase because larger-diameter pipes and additional pumps and pumping stations are required) (UNEP 2002b). It is common that wastewater master plans use economy of scale to justify increased sewage concentration and centralized, advanced systems. This, however, is not always the case. Changing numerous already paid-for, small, and easy-to-control systems into huge industrial-type structures, including large trunk sewers, pumping stations, and advanced treatment systems that require a high level of skill for their operation and maintenance, often only helps the designers, the big contractors, or the decision makers.

In general, it is difficult in developing countries to find budgets for financing clustered or centralized wastewater management infrastructure at the local level. Typically, when some financing does become available the central government often finances a scheme for a certain city or a group of cities, either through its own resources or, more likely, through international loans or donations. Exceptions to central government or international financing may be found in tourist areas, where the income from tourism has sometimes been sufficient to finance local clustered or centralized wastewater management systems.

Local Requirement 2: Effective Cost Recovery

For sustainable operation of wastewater management infrastructure, the annual recurrent costs may be even more important than the capital costs, since the latter are often funded centrally or from abroad. The local end-users or institutions need to be able to carry the O&M costs for sustainable operation.

Cost recovery versus ability to pay (affordability) is a current key issue in most developing countries. In some countries, it is argued that wastewater management services are expensive and should therefore be subsidized and a recurrent financing mechanism for intervention introduced. It is also emphasized that appropriate and affordable technology options are of overriding importance because, minor exceptions apart, wastewater authorities and companies have been unable to get community support for the works and thus have not received contributions toward cost offsets. For example, in South Africa it has been decided that full waterborne wastewater management systems should only be installed where residents are able to afford the full O&M costs of the system.

Whatever the level of operation and maintenance costs, resources have to be mobilized locally through municipal budgets, tariffs, recycling, or other income-generating activities. The most important aspect is that the locally determined financing method can cover the cost of O&M activities, thereby securing the long-term functionality and appearance of the wastewater treatment system. General local municipal budgets may be utilized, especially if the O&M costs are minor; however, if it is possible to create sources of income from the operation and earmark this income for O&M activities, the wastewater management system has a much better chance of becoming a sustainable component in the infrastructure of the community or municipality. Income generation, depending on size and type of system, may include selling the reclaimed water, selling the flowers and plants from the wetland units to private landowners, and connection fees. Wastewater fee collection schemes, in general, have not been a successful or viable method in most developing countries for financing of O&M costs.

Local Requirement 3: Decentralized Local Organization

Wastewater agencies are typically central or national organizations that are, traditionally, highly inefficient. Most developing countries still operate a unitary central system of government, with the national water and wastewater agencies having provincial offices to which varying degrees of power are delegated. In many cases, the management of these services in urban areas is conceded to a national utility, a parastatal corporation, or a private company jointly owned with the government (exceptions are countries where municipalities are given responsibility for providing the water and sanitation services in urban areas, such as South Africa and Ethiopia) (UNDP 2005).

These central agencies have typically been highly subsidized by the central governments; often employ many local people; are overstaffed with poorly motivated and poorly trained personnel; have inadequate equipment and technical expertise as well as meager financial resources; and are been afflicted with poor management practices. In general, there has been too large a proportion of

> **Box 3-15. Financing and Maintaining Local Wastewater Collection Systems**
>
> Financing the simplified sewer system for the urban poor in Brazil is based on households that pay for the on-site costs, blocks that pay for the block sewers, and water authorities or municipalities that pay for the trunk sewers. This simplified sewer system not only cut costs 20% to 30%, but also included the active involvement of the population in choosing their level of service, and in financing, operating, and maintaining the feeder infrastructure. The key elements are that families can choose to continue with their current sanitation system or to connect to the simplified sewer system. If a family chooses to connect to the simplified system, it has to pay a connection charge (which may be financed by the water authority) and a monthly tariff. Families are free to continue with their current system, which usually means a septic tank discharging into an open street drain. In most cases, however, those families who initially chose not to connect eventually end up connecting, either because they succumb to pressure from their neighbors or they find the buildup of wastewater in and around their houses intolerable. Individual households are responsible for maintaining the feeder sewers, with the formal authority maintaining only the trunk sewers. This increases the community's sense of responsibility for the system. Also, the misuse of any portion of the feeder system (e.g., by putting solid waste down a toilet) shows up as a blockage in a neighbor's portion of the sewer. The rapid, direct, and informed feedback to the misuser virtually eliminates the need to educate the users of the system in acceptable and unacceptable behavior, and results in fewer blockages than in conventional systems.
>
> The danger, however, is that the clever engineered system is seen as a final, almost automatic system. Where the community and organizational involvement has been missing, the technology has worked poorly, as in Joinville, Santa Catarina, or in Baixada Fluminense in Rio de Janeiro. The simplified sewer system in Brazil indicates the importance of creating a productive partnership between the communities and the municipality or the wastewater management authority (UNEP 2002c).

money and resources placed into these centralized institutional systems compared to local operational systems and activities. Central wastewater government institutions have kept control by means of regulations, central financing, central income collection, central definition of national health and environmental quality standards, personnel structures in the public service, and price structures. This situation has had dire consequences in relation to overregulation, inefficiency, lack of focus on O&M, and a disregard for on-site sanitation as an adequate and permanent solution in favor of completely new, centralized off-site systems. Sanitation authorities in most developing countries are self-proclaimed centralized wastewater authorities.

Some developing countries, however, have recently been engaged in a wastewater sector reform process, sometimes under pressure from multilateral and bilateral agencies, especially The World Bank and the regional development banks (UNDP 2005). The most important aspects of this modernization of the sector have been a move toward decentralization, municipalization, and in some cases, privatization. For example, in Chile the Regional Water and Sanitation Companies have decentralized and begun operating on a commercial basis; in Brazil, companies previously established in the states under the National Sanitation Plan have initiated management contract schemes following the central government policy; and in Colombia, greater responsibility has been given to municipalities for the development of infrastructure services, including drinking water and sanitation.

Decentralization within the wastewater sector is important for many reasons, including development of local organizational and human resources. Decentralization is required for the development of local capacities at the municipality or community level to maintain and operate the wastewater management system. The implementation of community-based cluster wastewater management systems may, for example, be carried out by local contractors who may in turn hire members of the local community as advisors to ensure knowledge of the local context, local suppliers, local pricing, loyalty, and mutual responsibility through local networks. Everyone develops local expertise in the functionality of the system, the facilities, and the individual installations—all issues of great importance for the continued O&M of the system. After completion, an "environmental fund" may be established, receiving income from sale of re-used wastewater, flowers and plants, and new wastewater connection fees, and managed by a local environmental committee, to ensure sustainable O&M.

Local commitment and accountability are always key issues for long-term sustainability. A crucial issue for the success of any wastewater management system is the group of people who benefit from it. Numerous wastewater schemes have failed completely because the designated local users and local financial supporters of the new infrastructure were not consulted about whether they valued the initiative and would be willing to contribute for its proper O&M. Thus, inadequate involvement of the local users and stakeholders during the planning phase created a situation of lack of demand.

For on-site systems, O&M activities are normally carried out by houseowners and private contractors who take care of desludging and disposal and treatment of sludge. Communal or municipal management of the whole on-site system may, however, in some locations prove to be a stronger setup.

> **Box 3-16. Local Community Latrine Management**
>
> Kibera, Nairobi's largest suburban area, has a population of 470,000. This slum has only 2,800 toilets, most not in good condition (170 persons per toilet). Drainage is virtually nonexistent. During rains, the area can hardly be traversed on foot and wastewater overflow is a particular nuisance. The Kenya Water for Health Organization helped residents to establish a latrine-emptying service, for which they were willing to pay in advance. With this help, the residents built ventilated improved pit (VIP) latrines and utilized a special suction truck able to maneuver through the narrow streets and empty the pit latrines regularly. A 13-member community management team was established to oversee the operation. More than 6,000 households paid the $9 USD advance fee to have their home latrines emptied. Sustainability was supported by responsibility-sharing between the community, the private sector, the public sector, and donors (UNEP 2002c).

3.9.2 Central Perspective

Moving from the individual or local system perspective to the central or national perspective, four issues are important. Accepting the principle that decentralization is fundamental for effective O&M of individual wastewater management systems, the key issue is how the central level can enable sufficient and effective local capacities. The four enabling issues concern institutional, financial, and legal setup, and political will. Each of these four enablers is discussed below.

Central Enabler 1: Enabling through Appropriate Institutional Setup

From Central to Local. Decentralization and deregulation are prerequisites for improved wastewater management. This includes decentralization and devolution of decision making to lower administrative levels; the right to raise money by, for example, tariffs; and allowing wastewater utilities to operate as autonomous entities so they can decide on tariff structures and personnel management. This also includes involving private partners to implement at least part of the management, financing, and O&M; to identify wastewater re-use rights and opportunities; and to apply financial (dis)incentives rather than inflexible command-and-control regulations to control, for example, wastewater discharges.

Decentralization and increased focus on expanding the access to wastewater management shifts emphasis toward local systems and influences household and collective action at the neighborhood level. This confronts the central wastewater management authorities with new challenges because progress requires that public agencies broaden their traditional service-provider role to include encouraging and supporting on-site and cluster systems. Decentralization also affects household and community actions, and collective decision making, by promoting solidarity, social capital, and the kind of hygienic culture that imparts value to improved on-site, cluster, or centralized wastewater facilities.

Most central wastewater agencies are unfamiliar with or are ill suited for this role. Wastewater management service agencies are typically modeled after utilities in industrialized countries and, as such, are organized around maximizing utilization and operational efficiency of centralized, large-scale wastewater management systems.

There is undoubtedly a need for governments to better define the roles of all institutions involved, to put in place mechanisms for coordination of key players, and to provide better frameworks for decentralization—with the central level providing facilitation and regulation, and the local municipal level providing management and O&M. The executive functions for large, centralized wastewater management development commonly reside in an engineering-based central government department or authority. By contrast, the executive functions of on-site and cluster wastewater management systems are often associated with urban management authorities that hold the mandate for land-use planning and housing regulations For example, they can force industries and workshops to move out of inhabited areas and into designated industrial zones, where the authorities are better equipped to separate and contain domestic and industrial wastewater flows. Unfortunately, most urban authorities show little interest in wastewater management, feel less accountable to national or regional environmental management, and typically limit their interventions to removing local wastewater to the border of their city. Regulatory functions are typically the responsibility of a national government ministry (health or environment). It is within these functions of government, among others, that changes are required to create an efficient, decentralized framework for wastewater management.

Locally, the organizational scale should reflect the scale of system—on-site, clustered, or centralized—and can therefore range from small, such as a city quarter or village, to large, the size of a metropolis. Because much O&M and cost recovery are physically associated with highly detailed wastewater collection networks and individual households, decentralization or devolution of responsibilities to the lowest appropriate administrative level is important. Part of the local network or infrastructure may be entrusted to a local water users' association or local community.

From Service Provider to Facilitator. What should be left for the central government is facilitation regarding promotion, public knowledge, technical capacity development, regulation, and establishment of mutual control among various agencies by creating watchdog organizations and balancing the power of one agency with that of another. Reorienting public institutions to broaden their influence on consumer behavior, as well as to engage community-level institutions in planning appropriate interventions, should be at the center of efforts to expand household access to private sanitation. For many countries, such a shift in strategy has major implications for central wastewater management institutions. For example, in policy and planning, the prevailing custom of linking sanitation exclusively with water supply must be reconsidered. Greater progress in expanding access to basic sanitation and provision of integrated wastewater management systems (which also includes re-use and re-entry systems) may result from forging strong linkages with other services that engage households in a more direct and continuous manner, such as health, education, agricultural extension, and rural development. Central authorities shall enable more strong roles of local government, community organizations, and small-scale private providers.

Central Enabler 2: Enabling through Sustainable Financial Setup
Presently, public resources are used within the wastewater sector for public investment in collective assets such as trunk sewers and wastewater treatment plants. Decentralization and the role change from service provider to facilitator, however, means that the most effective use of public funds may be in powerful marketing and promotion of sanitation and hygiene. Supporting ancillary services, such as microfinance, may also help local levels express and act on latent demand for service improvements, as well as to support an emerging pool of small-scale service providers who can respond to varied and changing demands at the local community and household levels.

A key factor of decentralized institutional capacity is the degree to which the service organization is financially autonomous and freed from the national budget. Authorities or agencies that derive the bulk of their revenues from user payments such as water and sewer fees, connection charges, or special taxes are also the most stable. This is seen in Société Nationale d'Exploitation et de Distribution des Eaux (SONEDE) in Tunisia, Régie de Distribution d'Eau (REGIDESO)in the Democratic Republic of Congo, Société de Distribution d'Eau de la Côte d'Ivoire (SODECI) in Cote d'Ivoire, West Africa, the Water Supply Department in Nairobi, Kenya, the municipalities of South Africa, the transformed Water Corporations in Nigeria, and in Patong and Pattaya municipalities in Thailand.

The acute shortage of funding for clearing the backlog and expanding the wastewater services as urban populations continue to explode has also led some countries to develop alternative financing strategies. One includes tapping the private sector's resources. This approach has had rather mixed success because it depends on the political will and purpose of the privatization; for the most part, this been unsuccessful when applied to larger centralized systems.

Lack of access to credit may impede investment in wastewater management, especially in small cities and towns. This problem has been overcome in some cases by creating special municipal development funds or rotating funds to finance environmental investments. Poor urban households need mechanisms to finance sewer connections and in-home sanitary facilities, and some cities provide credit to poor households for these investments that can be paid off in installments over periods of 3 to 5 years. The installment payments may be collected as part of the monthly water bill. In some cases, households can provide sweat equity (labor inputs provided by the community for self-help construction schemes) or even make partial payment in the form of construction materials.

A special sanitation credit fund has been established in Honduras for poor urban households, fashioned along the lines of the well-known Grameen rural credit bank in Bangladesh. Such experiences show that the urban poor will invest in improved wastewater management if they can spread the initial costs over time. Similarly, innovative schemes for providing urban households access to credit for sanitation investments have been demonstrated in Lesotho (southern Africa), in Burkina Faso (West Africa), in Brazil (where the World Bank has supported the creation of municipal development funds in the state of Minas Gerais for environmental improvements in small cities and towns), and in Mexico for municipal water supply, sewer, and solid waste investments in intermediate-sized cities (UNDP 2005).

Central Enabler 3: Enabling though Appropriate Regulation

For many industrialized countries, the approach has been to set universal environmental standards and then raise the funds necessary to finance the required investments. It is becoming increasingly evident that such an approach is proving to be very expensive and not financially feasible, even in the richest countries of the world (as a case in point, about half of all wastewater in France is not even being treated, let alone reaching the high EU discharge standards). Regulation, instead of being based on general standards, may, for example, be catchment- or local community-based, where people in a certain basin or pollution-sensitive area are involved in setting standards, in making trade-offs between cost and ecosystem improvements, and in ensuring that available resources are spent on those investments that yield the highest environmental return. They may apply financial incentives to encourage users and polluters to reduce the adverse environmental impacts of their activities. Effluent taxation is one obvious form of incentive that is used in many countries, as it induces waste reduction, encourages treatment, and can provide a source of revenue for financing wastewater treatment investments. Effluent taxation, however, requires a high level of institutional and organizational resources which normally are not available in most developing country settings.

Most legislation and regulations pertaining to wastewater management in developing countries are sketchy, uncoordinated, and sometimes conflicting, and most existing laws adopt a generalized approach with specific details left out. Often, too much emphasis has been given to improving and detailing the central-level legal instruments, resulting only in strengthening the legal power base of the central level when emphasis should be given to improving the decentralized approach.

Central Enabler 4: Enabling through Political Will and Stability
Lack of political will, or political instability, may be the most serious, immediate constraint on the improvement of urban wastewater management services in developing countries. Lack of political stability and will to decentralize, create efficient institutions and legal frameworks, support the central facilitator role, and minimize disrupting intervention in tariff and other decisions vital for sustainability of the services makes it difficult to plan ahead, maintain implementation schedules, or create stable and efficient local wastewater management systems. Political consensus regarding the overall short- and long-term national goals for the wastewater management sector (e.g., decentralization or increased focus on on-site and cluster systems) is fundamental for actual improvements in the local delivery of wastewater management services.

))) 3.10 Nodding in Perspective: The Width and Depth of Assessing Appropriateness and Sustainability

We have now described the 10 guidelines for appropriateness and sustainability, stressed the importance of contextual understanding and fitness, introduced the six elements of appropriateness of wastewater management systems, and provided examples of scales and technologies. Of course, we would nod approvingly if we saw that all of this has been incorporated into the design and implementation of a specific appropriate and sustainable wastewater management system. But that would, in most situations, be unrealistic. What is important is to maximize what is possible and practical in an actual situation. We therefore also nod when we see an appropriate and best possible mix and match of the six elements for good wastewater management.

3.10.1 The Invisible Checklists: The Width of Assessing Appropriateness and Sustainability

Even though we generally do not like checklists, we have developed two more: the "six-element" and the "smart technology" nod checklists. When we started this book we sat down and discussed our experiences. Out of that came our "nod lists." However, it is not the lists but, rather, the thoughts and considerations they stimulate, that are important, so we therefore would like to call them the "invisible nod checklists." Checklists help us to not forget the big picture, and, if applied seriously, to be honest (Fig. 3-10).

The nods work as design criteria or guidelines in the planning process and as checklists upon which the level of appropriateness and sustainability can be evaluated. The more nods in a management system, the more likely it will succeed as an efficient and site-optimized response to wastewater treatment challenges in the context of developing countries.

The different elements and the multiple possibilities of mixing and matching the elements and technologies are presented in case studies in the coming chapters. These cases are interesting because they incorporate one, several, or many of the principles and elements in the nod checklists, and they present different mixes and matches of the six elements and of different technologies. After each case presentation, we will reflect on the overall appropriateness and sustainability of the case, or lessons learned, followed by details and discussion of interesting technologies applied in the case. We will apply our invisible nod checklists to each case.

3.10.2 The Invisible Checklists: The Depth of Assessing Appropriateness and Sustainability

As with any checklist, applying the invisible checklist has both a width and a depth. The *width* for appropriate and sustainable wastewater management

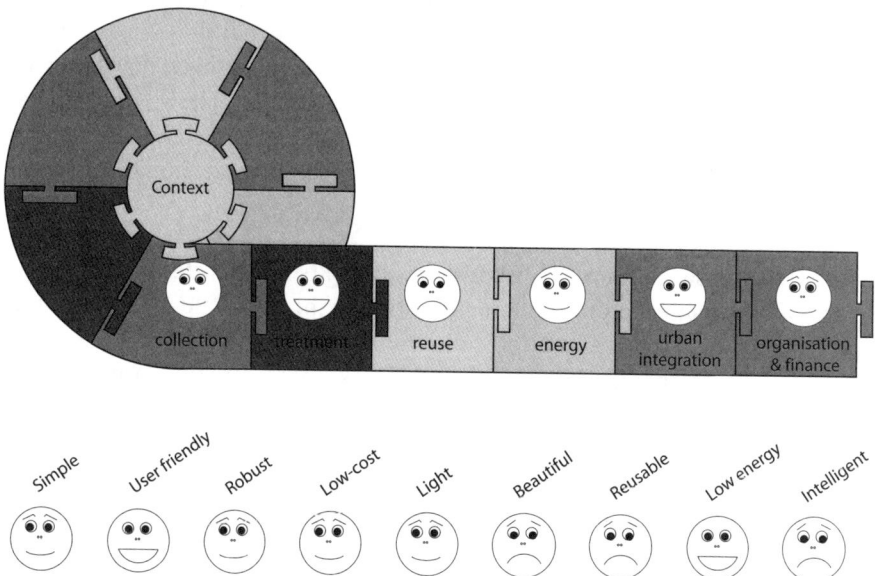

Figure 3-10. The "invisible" six-element and smart technology checklists.

is the six elements (collection, treatment, energy consumption, urban integration, re-use and re-entry, and organization and finance) across the board. They are spottable; they can be seen and ticked off: Have they been applied?—yes, no, or only partially.

But good wastewater management systems are not a simple question of ticking of a yes, no, or partial box. It is a question of being the best and most appropriate system for this problem at this location at this time, with these people with their competencies, interests, motivations, and resources. This means that the most appropriate and sustainable system would include important choices and judgments. In a specific wastewater management system, should we include all elements? If not, which should be included and which left out? For example, do we really have to accept that wastewater cannot be re-used in a certain situation? Have all alternatives been carefully considered before we gave up on a certain element? Or was it acceptable that this element was not included? But not only that: Within each element we must decide on the multiple alternative parameters, solutions, technologies, and approaches available. The choices are many and interlinked. This is the complex matter of *depth* in appropriateness and sustainability, in the choices of planning and design, and in the contextual assessment.

Is a specific wastewater management system appropriate? Actually, that depends on the eye of the beholder. It depends on our ability to assess, and our ability to assess, design, plan, and implement depends on our competence and practical experience. Everybody does not assess, design, plan, or implement equally well! It is basically a question of one's level of professionalism. But then, what is professionalism and professional capacity?

An interesting answer to this important question has been developed by Hubert and Stuart Dreyfus (1986) that with great clarity shows the qualitative depth aspects of professional capacities. These two researchers show that we go through five different phases when we develop professional capacities (Fig. 3-11).

Level 1: Professional Capacity Based on Context-Independent Rules

At this level, we meet a problem or situation for the first time. By instruction and training we learn to recognize different objective facts and characteristics of the situation, and rules are taught. Facts, characteristics, and rules are defined so clearly that they can be recognized without being related to the specific, concrete situation they operate within. They can be generalized for all similar situations. They are context-independent. On this level we value and are valued based on how well we follow the rules we have learned. When we have learned a handful of rules, execution becomes so complex and demands so much concentration that this in itself will limit further improvements in our ability to act. These first rules are necessary to get the first experiences,

> **Box 3-17. The Dreyfus Experiment**
>
> In the mid-1980s Hubert Dreyfus, a professor of philosophy at the University of California–Berkeley, and Stuart Dreyfus, a professor emeritus in that university's Department of Industrial Engineering and Operations Research, developed their Model of Skills Acquisition. This model describes how learners progress through five distinct stages of learning: the *novice*, who wants recipes, best practices, and quick wins; the *advanced beginner*, who wants guidelines and a safe environment in which to make mistakes; the *competent* stage, where one wants goals and the freedom to execute; the *proficient learner*, who wants maxims, war stories, and metaphors; and the *expert*, who wants philosophies, discussions, and arguments with other experts.
>
> The researchers experimented with a group of healthcare employees. Six persons were videotaped independently while they were resuscitating patients through heart massage and artificial respiration. Five of the six were inexperienced students currently being trained in life recovery. The sixth was a professional healthcare employee with solid experience in resuscitation. These videos were shown to three different groups: healthcare workers with practical experience in resuscitation; teachers in resuscitation; and students in this discipline. Each spectator was asked the following question: Which of the six persons on the videos would you choose to resuscitate yourself, if you had had an accident? Ninety percent of the experienced healthcare workers chose the experienced healthcare worker. Fifty percent of the students correctly chose the experienced healthcare worker and, surprisingly, only 30% of the teachers got it right (Dreyfus and Dreyfus 1986).
>
> Interesting, to say the least. What made the teachers perform so poorly and why did the experienced healthcare workers choose so correctly? The key, which Dreyfus and Dreyfus showed so effectively, is experience—having done the same thing over and over; having learned from numerous real-life experiences and cases.

but these rules quickly become a hindrance in the learning process and must be set aside in order to move on. *The invisible checklists have been read, studied, and understood but not experienced.*

Level 2: Professional Capacity Based on Context-Independent Rules with Some Added Context-Dependent Elements

When we have seen a problem or situation a few times, we recognize them as analogous to earlier similar situations. This brings us to the next level on the professional capacity learning curve, where actions are done in a more nonreflective and automatic manner, not only by using context-independent facts and rules. By gaining experience from real life, we advance from the theoretical and protected situations at the first level. Through these experiences

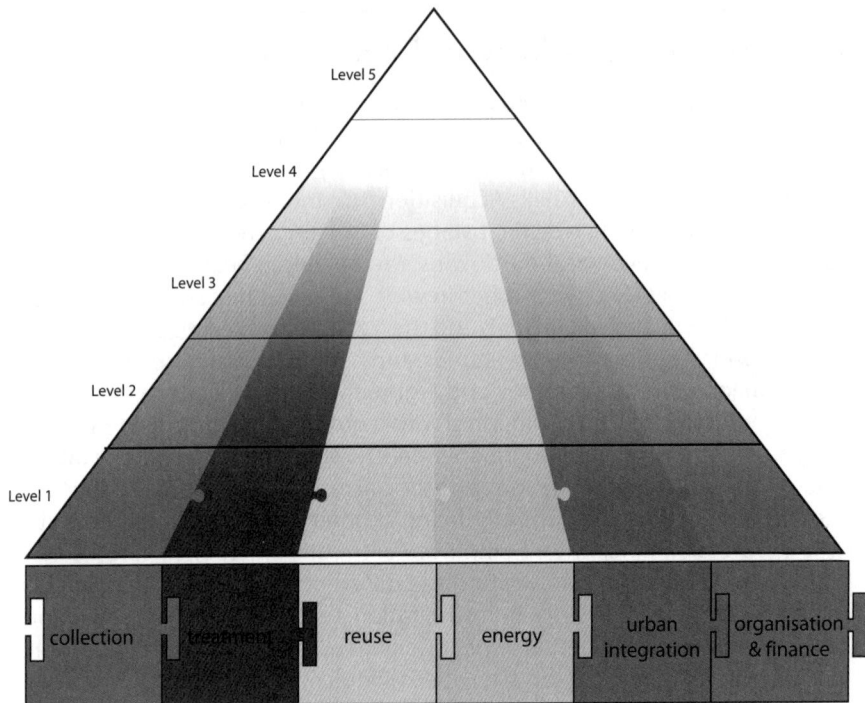

Figure 3-11. The width and depth of professional wastewater management capacity.

we start to recognize relevant elements in relevant situations. The recognition is concrete and dependent on context. On this level, rules can therefore be both context-dependent and context-independent. Real-life experience (e.g., trial and error) is on this and the next levels more important than any context-independent and explicit formulated facts and rules. Context becomes increasingly important. *The invisible checklists have been read, studied, understood, and seen applied a couple of times but have not been used for personal decision making.*

Level 3: Professional Capacity Based on Goals and Plans—the Basis for Involved Actions

With increased experience, the number of recognizable elements we can recognize in a specific, real-life situation becomes very large. However, on this third level we still lack an understanding of which elements are important. We lack the ability to prioritize. On this level we are taught to apply a hierarchical, prioritizing procedure for decision making. By choosing a goal and a plan, by organizing the information of the specific situation, and by only dealing with the important and relevant factors we can both simplify and improve our results and achievements. Goals, plans, and prioritizing make us deal only with a limited set of important factors instead of having to deal with the combined and total knowledge of a given context-specific situation. We are now beginning to get better adapted to the specific context. We are becoming the competent doer.

To choose a plan on this level, however, is not simple and without problems. It takes time and is done consciously and carefully. On this level we do not have any objective, rational procedures for our choice of plan, as we had on the first levels—our context-independent selection of facts and our use of rules. Our choice of plan will, furthermore, have extensive consequences for our actions and results. The lack of fixed points for our choice of plan, combined with the necessity to in fact have a plan, results in a new, important issue: involvement.

At the first two levels we experienced only limited responsibility for the results of our actions. We used specific elements and prescribed rules to undertake actions. Thus, a bad result would, if we not had made a blatant mistake, appear to be the consequence of insufficient rules. On this level, however, this does not hold true any more. On this level, after having struggled with the problem of choice of plan, we feel responsible for the consequences of our choice because that choice was not made objectively. The important elements of *interpretation* and *judgment* influence our actions on this level; according to Dreyfus and Dreyfus, interpretation and judgment are at the core of true human capacity and expertise. *Our invisible checklists have thus been read, studied, understood, experienced, and been used as the basis for planning, designing, and decision making on a number of actual implemented wastewater management systems.*

Experienced doers, or experts, display a quick, intuitive, holistic, interpretative, and visual mode of thinking and action—quite dissimilar to the slow, analytical mode of thinking that characterizes rational problem solving at the first two levels. One-sided focus on analytical rationality limits our best achievements because of its slowness and focus on rules, principles, and universal solutions. Speed and a thorough knowledge of specific cases is a precondition for true capacity and expertise. The ability to make interpretations and judgments becomes even more crucial at the remaining two levels.

Level 4: Professional Capacity Based on Context-Dependent Intuition and Experiences, but with Added Analytical Problem Assessment before Action

So far, even if we have progressed beyond just abiding by rules and prescriptions, we have still only chosen goals, made decisions, and taken actions after deliberate reflection on different alternatives. Compared to this, the decision-making process on the fourth level is more fluid and less phased in time. On this level we are typically deeply involved in our actions and have developed a *perspective* based on earlier situations and experiences. This perspective makes certain aspects of a situation stand out more clearly and distinctly, while others will be more blurred, indistinct, and less important. New situations and experiences will change the aspects that stand out, change plans and expectations, and thereby influence actions.

Here, we do not make a purely rational choice of aims and plans or a conscious assessment of problems and solutions. Our choices, assessments, and judgments are obviously made but are based on our earlier experience with similar situations. We understand and organize our tasks intuitively but still occasionally think analytically about what should happen. Intuitively, certain elements and plans stand out as important and relevant, and we assess them and combine them analytically, with the help of rules, to arrive at the most appropriate decisions. Our deep, intuitive involvement interchanges with analytical decision making. *Plans, designs, and decisions are made and only occasionally checked with the invisible checklists.*

Level 5: Professional Capacity Based on Intuitive, Holistic, and Synchronic Problem Assessment and Action

According to Dreyfus and Dreyfus, this final level is not is not reached by everyone. This is where situations are recognized intuitively and the relevant decisions, strategies, and actions are judged and acted upon intuitively, synchronically, coherently, and comprehensively. This is the level of true human capacity and expertise, and is characterized by fluid, free achievement. This is the level of virtuosity. The virtuoso does not see problems as one thing and solutions as something else. The virtuoso does not even make plans; he or she just does it. Dreyfus and Dreyfus equate virtuosity with intuition; others have added traits like creativity and innovation. *The invisible checklists have been more or less forgotten!*

3.10.3 Five Levels of Depth of Capacity—So What?

These five levels of depth of professional capacity provide us with several important insights: Why it is so important to not use checklists blindfolded. Why some people will do better than others. Why experience, and teams with different types of experience, are so important. Why some systems are designed to *not* work, because they are designed by people with Level 1 professional experience!

Apart from the obvious rational prescriptions, rules, and checklists, professionals working within the wastewater management sector need to develop context, praxis, trial-and-error experience, common sense, intuition, creativity, and innovation.

At Level 3 an important change in our professional capacity occurs: The most important basis for action is no longer analytical rule thinking, but context, experience, and intuition. Action based on logic is superseded by experience-based action. The five-level model helps us remember that analytical rationality is not all there is; the latter does not show us the full spectrum of professional capacities. Analytic, rule-based rationality mainly focuses on where we mostly act as inexperienced engineers or bureaucrats, on the first and second levels of professional capacity.

This is not to say that analytic, rule-based rationality is unimportant. The first two levels provide a basis—things and approaches we must know. This book will hopefully provide a part of that basis of things we should know when working with wastewater management in developing countries.

))) 3.11 Sense and Simplicity

Mixing and using different but complementary wastewater management systems, elements, and technologies is easier said than done when abundant competing approaches exist on the market. Our 10 guiding principles for appropriateness and sustainability provide some guidance, and a little structure in the many approaches can be achieved by using the framework of the six elements. Having knowledge of different but complementary wastewater management systems is one thing; it is quite another to apply these in real-life settings, where planners and decision makers must balance, choose, and mix different systems, elements, and technologies.

Developing an appropriate wastewater management system is about pinpointing the most important contextual issues, making balanced judgments, and then choosing, designing, implementing, and continuously readjusting along the way. Imagine the final meeting of a wastewater system design team: Elements have been chosen, judgments made, technologies mixed and balanced, and the team is ready to go ahead with detailed design and rendering. This is one of the most important times to rethink and reconsider, and we have found that at this stage it is useful to use the following two tests before moving ahead.

3.11.1 The "Does-It-Make-Sense" Test

First, do the "does it make sense" test. Because this is so crucial a factor for success and sustainability, and because planners can so easily be caught in their own ways of thinking and justifying, it is useful, one final time, to ask and reflect upon: Does it make sense? The tricky issue here is that it almost always makes sense for someone—for the central governments providing the finance; for the engineer who is an expert in activated sludge; or for the coastal ecosystem environmentalist. Nevertheless, today's numerous malfunctioning treatment systems indicate a lack of making sense in terms of who is supposed to continue operating, maintaining, and financing the system.

In this test, consider two key target groups: first, does it make sense to the mayor, the municipality director, or the head of the local community—the one who is actually to become responsible for financing, operation, and maintenance? Will they justify the money, resources, and inputs necessary to make the wastewater management system work? Second, does it make sense to the normal citizen in the area? Will he or she, with their income, work situation, and education level, consider the required money, resources, and inputs justifiable compared to the improvements achieved and to alternative usages of these resources and finances? If the answer still is yes, it probably will make sense to the local decision maker and citizen; sustainability, positive impact, and robustness are more likely; and the implementation process may proceed.

3.11.2 The Simplicity Test

The second test is the simplicity test. Is the proposed system designed as simply as it could be? Of the many factors defining appropriate and sustainable wastewater management systems, simplicity may be the most important and useful. Others could also be used (e.g., robustness, low energy consumption) but we have found that a focus on simplicity can most often help improve a system's chances for survival.

The starting premise is that, because creating wastewater management systems that actually work is already very complicated, there is no need to add unnecessary technical complications. Therefore, wherever possible, system complexity should be reduced. Is pond treatment sufficient? Could the number of pumps be reduced? Is this component necessary? Could it be made simpler? Could the intake structure be built more simply with fewer mechanical components? Every time process complexity is reduced, the need for maintenance, the cost for replacements, and the required technical operational knowledge are also reduced, improving the chances for system sustainability.

4

Sustainable Wastewater Management at the Chairman's House: A Recovery-Based, Closed-Loop Household System

))) 4.1 The Living Lab of Dr. Ksemsan Suwarnarat

The former chairman of the Wastewater Management Authority of Thailand, Dr. Ksemsan Suwarnarat, has a lifetime of professional experience in wastewater management and a wholehearted personal commitment to the task of developing and implementing appropriate wastewater management systems. He has made it his hobby to use his private home as a living laboratory for experiments with these systems and technologies (Fig. 4-1).

Based on the concept of zero discharge, no stormwater, wastewater, sludge, or organic wastes from the household kitchen and garden are allowed to leave his private plot. Even batteries are integrated into his re-use system. This zero-discharge approach eliminates his household's need for municipal drainage or wastewater collection systems and minimizes the need for an external water supply, soil fertilizers, and solid waste collection systems. Let us take an eye-opening tour through the living laboratory of the Chairman's house.

Use the (b)rain: the soil is a sponge. All rainwater is collected from the roofs and is discharged into cisterns, from where the water seeps into the soil. Not only does this make the trees and vegetation in the garden lush, it also keeps the soil from drying out and saline waters from entering the system. Besides, there is no discharge to public drainage systems and minimal demand for additional irrigation.

From atop his studio apartment on the premises, the Chairman collects the rainwater from the roof and uses it in the kitchen, for showering, and for toilet flushing. This apartment is entirely self-sufficient by utilizing rainwater and has no connection to public waterworks.

Figure 4-1. Collage of the home of Dr. Ksemsan Suwarnarat.

Everything to the drain, batteries included! The studio apartment has another feature, an aqua privy. For more than 30 years Dr. Ksemsan has advocated aqua privies as the most viable solution to on-site waste treatment. Consequently, the apartment has a septic tank directly beneath the toilet, which receives all organic wastes from the toilet and kitchen. The system works as an anaerobic tank in which the biological waste decomposes. During the process, large amounts of hydrogen sulphide are released, which is utilized to bind the polluting heavy metals from batteries, so in this apartment the batteries are flushed into the toilet as well! The privy then works more or less as a septic tank, with floating and settled sludge and a liquid effluent that, like the rainwater, seeps into the garden soil (Fig. 4-2).

Flush your wastewater and harvest the fruits. A total of eight people are contributing to the Chairman's third experiment—an anaerobic tank where sanitary and kitchen wastewater is digested followed by a subsurface irrigation system supplying water to the herb and heliconia ("False Bird of Paradise")

Figure 4-2. The drain field.

garden of the household. The anaerobic tank is filled with small plastic balls with uneven surfaces developed to collect and encapsulate gases vented from the decomposition process and increase the efficiency of the tank. These balls represent another concept developed and patented by the Chairman. The effluent is discharged to the garden through an underground distribution system. Thanks to new inexpensive, perforated pipes, which can be bent in all directions, the irrigation water from the anaerobic tank can easily be distributed into the soil matrix. Some of the water and the rich content of nutrients is taken up into the plant cells; some water evaporates into the air through plant transpiration; and the remaining water seeps into the soil and recharges underground freshwater sources. All wastewater is utilized. The plants irrigated by the effluent from the anaerobic treatment provide flowers, salads, and fruits for the Chairman and his family.

Sludge used as fertilizer. To emphasize the principle of zero discharge, sludge is emptied from the anaerobic tank every 1 to 2 months and is spread on the soil as fertilizer (Fig. 4-3). Because the sludge has been fully decomposed during the anaerobic treatment, there is no odor problem. Applying the nutrient-rich sludge as soil fertilizer, especially around the bigger trees in the garden, eliminates the need for municipal sludge collection, sludge drying, and a sludge disposal system. No disposal or landfill is required because all surplus sludge is managed and re-used within the garden.

Not just hot air. As an experiment, a part of the effluent from the anaerobic treatment is diverted into an aerobic treatment unit. The aerobic digester is designed as a rotating biological contactor (RBC) where microorganisms settle on a wheel that slowly rotates in and out of the wastewater, thus partly exposing the bacteria to air and partly submerging them in water (Fig. 4-4). The wheel is rotated by a small, mass-produced, inexpensive motor that consumes about the same amount of energy as a medium-sized light bulb. The effluent from the aerobic treatment unit is clear and odor-free, and is

Figure 4-3. Using sludge as fertilizer.

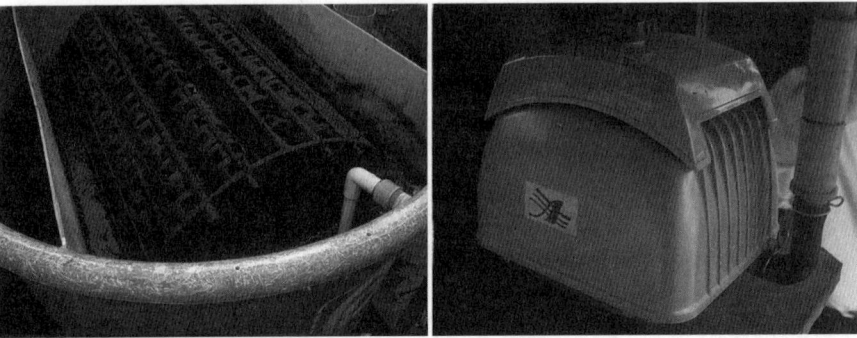

Figure 4-4. The rotating biological contactor (RBC) and pump.

discharged into a small, open pond in the garden. (Chapter 7 contains a more detailed description of RBCs.) From there it flows along a canal at the edge of the lawn, irrigating the numerous colorful flowers in the garden.

Piece of cake. Fully aware of the importance of grease traps in Thailand, where relatively large quantities of oil are used to prepare the delicious Thai food, the Chairman has installed a small basket-like unit on the pipe from the kitchen. This traps oils and greasy items, preventing them from entering and possibly clogging the pipes and treatment units farther downstream. One day, watching his wife preparing a cake, Dr. Ksemsan came up with a simple idea for how to manage the greases trapped in the basket. He collects the greases and puts them in a cup with a small hole in the bottom. The cup is placed in the garden so the water seeps out of the cup and into the soil while the greases turn into a solid cake, which is either distributed in the garden as soil conditioner or discharged to the heart of the Chairman's organic solid-waste management system—the compost tank.

))) 4.2 Reflections on Appropriateness and Sustainability

The six-element checklist described in Chapter 3 indicates whether all elements in an appropriate and sustainable cyclic wastewater management system have been dealt with, and also roughly to what extent each element is considered to fit the local setting. As can be seen in Fig. 4-5, Dr. Ksemsan's on-site system scored high on all six elements in the management system, suggesting that is appropriate for smaller households, at least, but also probably for many other situations.

Also, three integrated reasons confirm that Dr. Ksemsan's system is appropriate and sustainable, namely simplicity, locally taking care of the problem, and the use of common, already available technologies.

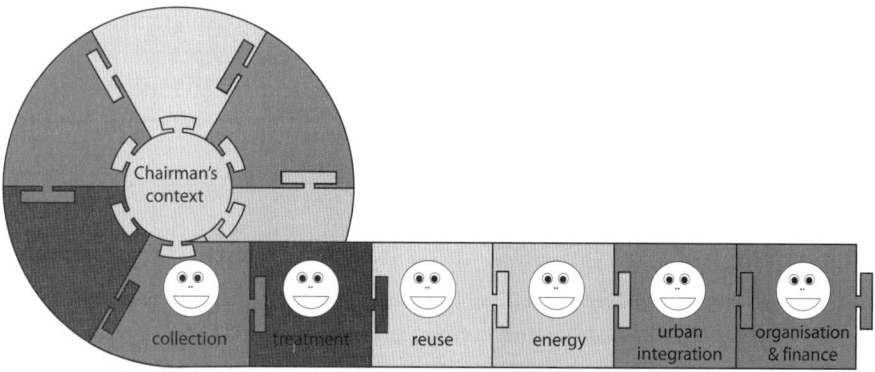

Figure 4-5. Contextual appropriateness scoring of the six elements of the wastewater management system at the Chairman's house.

Smile, contextually appropriate; no expression, somewhat appropriate; frown, not appropriate.

It is simple. The system of a septic tank collecting grey and black wastewater, and being connected to a drainage pipe distributes the effluent from the septic tank onto the roots of the plants in garden, make the wastewater management system at the Chairman's house highly appropriate. It solves the problem; it gets rid of the wastewater in a hygienic and environmentally safe way; and it does so in a very simple way. A pipe from the kitchen, an underground tank, and a few meters of perforated pipe are all it takes. Despite the simplicity, the system covers the whole spectrum of collection, treatment, reuse, energy, integration, and organization and financing.

The system does not need an extended collection system and the treatment takes place without human involvement or energy supply (not including the aerobic experiment). The water is re-used for irrigation; the system fits the building typology of a single-family house, thus being an example of appropriate urban integration; and the system has been implemented at half the price of a medium-sized TV. All in all, the system is a recovery-based, closed-loop household system that fulfils the task of managing wastewater in a simple but efficient and sustainable way.

It locally solves the problem. The wastewater management system at the Chairman's house applies in practice the rule of thumb that the level of complexity of a system should match the problem it has to solve: the small amount of wastewater produced from a single extended-family house. Why should wastewater from this household be sent, by large concrete pipes and at high expense, far away to be treated at a large wastewater treatment plant that needs large and continuous public expenses for staff, electricity, and O&M? Why should the problem be solved with a level of complexity that far exceeds the scale of the problem?

The biggest advantage of an on-site wastewater management system is that no additional systems need to be applied—no sewers, no treatment plants, no public investments, no O&M costs, no external dependency or burden on municipal administration and manpower, no tax collection, and no large-scale infrastructures to be maintained.

The Chairman's house represents a closed-loop system where water and nutrients are integrated in a holistic, cyclic, and self-reliant logic. Reusing wastewater within the premises of the single household unit enables the residents to save money because the water bill is reduced, the purchase of soil fertilizer is unnecessary, and there is no optional discharge fee to public drains and sewers. But perhaps more important is that the flowers bloom, the vegetables thrive, and the trees grow.

It uses common, already available technologies. On-site waste treatment at individual plots, household estates, commercial complexes, and factories is a legal requirement in most developing countries. For many years this policy has been successfully implemented and enforced in Thailand, which means that on-site wastewater treatment facilities are already widely implemented throughout the country. The Chairman supports the policy of on-site treatment and utilizes the legislation that has already been successfully implemented.

Installing septic tanks at the household level distributes the investment costs from the central administration to the private landowners. The cost of the tank is integrated into the overall construction cost of a new house, as important as the roof, the kitchen, and the garage—it is not a prohibitively expensive additional burden. The successful template the Chairman has produced should be widely repeated. *On-site treatment is most appropriate for most areas.* On-site systems are appropriate and applicable to the large majority of urban households and all rural settlements. It is estimated that such on-site wastewater management systems can benefit as many as 80% to 90% of all households in developing countries, indicating the potentials and relevance of pursuing the continued application and improvements of such systems. If all households had the same system as Dr. Ksemsan, there would in fact be very few problems with wastewater.

Specific issues for dense areas. For households located in areas with limited land availability and a very high population density, the concept of a recovery-based, closed-loop household wastewater management might not be suitable. Collection systems and municipal treatment facilities might have to be implemented in such areas. However, with continuous improvements and innovations, on-site systems should not automatically be disregarded for more dense areas. If the same resources were put into the development of on-site wastewater management systems as for centralized systems, these systems might also be made efficient for more dense areas. New, innovative on-site designs have, for example, been developed for high-rise buildings, with wastewater being treated in hanging gardens on the walls of the buildings. James Wines of New York-based SITE Architecture, Ken Yeang (T.R. Hamzah & Yeang International) in Malaysia, John Todd Ecological Design in Massachusetts, and architect William McDonough in Virginia have worked on some of these ideas.

Specific issues in groundwater-sensitive areas. A concern regarding seepage systems is the risk of contaminating groundwater sources. If it is not possible to implement seepage systems at least 1 m above the groundwater level, so that possible contaminants can be adsorbed in the soil matrix before the effluent reaches the groundwater, on-site systems will not perform efficiently. The risk of groundwater contamination, the actual state of the groundwater, and the present utilization of groundwater must be assessed carefully as part the development of an on-site "dilution is the solution" strategy.

Specific issues of awareness and system support. Dr. Ksemsan's integrated on-site system works in the private home of one of the leading experts in wastewater management in Thailand. However, what about at the home of his neighbor, who has no interest or training in the field and might not be either able or willing to spend the time and costs for installing, operating, and maintaining such a system? The tasks of, for example, collecting grease-cakes and manually emptying sludge from the septic tank and pouring it on the trees might require more active participation than many people are willing to commit to. City-wide implementation of an extensive zero-discharge system like the one at the Chairman's house would have to address uncertainties of the actual scale and efficiency of the wastewater management system. For example, some households might never empty their septic tanks, and others might discharge effluent from their kitchens onto their sloping front yards, accidentally creating a flow to the public street.

Once the wastewater management system is based on the active collaboration of people and managed on private land, the system must first make sense to the local people and there must be an individual benefit by choosing and contributing to the system. A simple septic tank connected to a sub-surface irrigation system fulfils these criteria of making sense and providing benefits. Second, the efficiency of on-site wastewater management systems should be enhanced by municipal assistance whereby professionals from the local administration support, check, clean, and rehabilitate on-site wastewater management systems on a regular basis. The O&M of on-site systems could be a public service provided for the residents. Compared to the expenses incurred by wastewater management systems with municipal collection and treatment facilities, the public authorities could in fact provide assistance free of charge for many years before the expenses come even close to the investments required for city-wide, centralized systems.

))) 4.3 Smart Technologies at the Chairman's House

The wastewater management system at the Chairman's house contains a number of appropriate and sustainable technologies. One of the most interesting and promising for on-site wastewater management systems in developing countries is the approach that combines septic tanks with subsurface irrigation systems (Fig. 4-6).

4.3.1 Septic Tank Combined with Subsurface Irrigation

The smart technology-spotting ace would see this as an elegant way of treating and utilizing wastewater. It is beautiful, it is underground, it fits well into the landscape, and it is simple and natural. Also, the checklist scores the technology high on most of the elements, indicating that the technology is a promising smart technology for wastewater management systems in developing countries. Simplicity, user friendliness, low cost, robustness, ease of operation and maintenance, and low energy consumption are important issues for wastewater management in developing countries, and it is evident that this technology positively addresses all of these issues.

Keeping in mind that we tend to nod approvingly when something passes the simplicity test (when things are not made more complex than necessary), a septic tank connected to a perforated pipe is a simple and robust technology efficiently solving the problem at hand. In addition, it creates more beautiful and green surroundings and is tailored to fit the exact amount of wastewater produced in the household without being overdimensioned or unnecessarily expensive.

4.3.1.1 The Technology: Septic Tank and Local Subsurface Irrigation

The septic tank is a simple cement structure equipped with tees at the inlet and outlet. The submerged inlet and outlet allows for separation of particles and water. Floating particles are collected in the scum at the top of the tank and settleable particles are collected as sludge at the bottom. The treat-

Figure 4-6. Scoring of each the nine elements defining smart technologies for the septic tank and subsurface irrigation technology at the Chairman's house.

Smile, supportive element for overall potential; no expression, somewhat supportive; frown, not supportive.

ment is based on basic principles of hydrology, gravity, and density, and it operates without an external energy supply.

The main benefit is that it is a simple and robust system that rarely breaks down. In addition, it is a well-proven technology that has been implemented in all parts of the world for many years. Thus, suppliers and contractors familiar with the technology can be found everywhere, which is important to ensure sustainability of the system. Finally, the septic tank has a very low level of sludge accumulation, which means that the tank can operate without manual interference for many years. Such an on-site system, due to low usage of toilet paper and the high temperatures inside the septic tank, can function maintenance-free for up to 10 years before the tank is filled up with sludge, eventually clogging the toilets; then sludge is removed and the septic tank can go on working for many more years.

In the Chairman's system, the water flows out of the septic tank and into an irrigation system in the garden by way of a perforated peak-hour pipe. The design of the outlet utilizes the high flow of water produced in the morning, when everyone in the family queues in front of the washroom, as the continuous hydraulic load generates a small current out of the septic tank and into the pipe. The perforations on the bottom of the pipe allow some of the water seep out of the pipe and into the surrounding soil, thus irrigating and fertilizing the garden environment.

The irrigation system, or slow-rate seepage system, uses the soil's filtering capacity to convert treated wastewater into a valuable plant asset. The garden plants have been added according to the quantity of water produced by the household and distributed into the garden, and through experiments with the type and number of plants that fit the moist of the soil. The majority of the water taken up by the plants is vented into the air through evapotranspiration, while some other becomes a crucial part of the plant cells. The slow-rate seepage system has no public health hazard because the drip-irrigation takes place underground. No raw wastewater is exposed to humans and the water is not in direct contact with digestible plant leaves. The water is taken up by the roots and is naturally and safely adsorbed in the plant.

4.3.1.2 Technical Considerations, Balances, and Choices

Two technical issues are raised by the system at the Chairman's house, namely, the anaerobic system versus the combined anaerobic and aerobic system, and the distribution of wastewater to the irrigation and seepage area.

Septic tank versus septic tank combined with a rotating biological contactor (RBC). The septic tank treats the wastewater in an anaerobic environment and requires nearly no maintenance, but it does produce a foul-smelling, black-colored effluent not suitable for exposure. In comparison, the extended

RBC system provides treated water that is clear and usually not odiferous, but it does require energy to keep the small, rotating wheel in motion. Sludge must be emptied regularly to maintain treatment efficiency.

The choice between an anaerobic septic tank and an extended anaerobic/aerobic system should therefore be made according to how the effluent will be handled, and the maintenance requirements. If the effluent will be sent to an open water surface that is integrated into the layout of the garden, the higher level of treatment with an RBC is preferable. However, if wastewater is to be discharged subsurface and minimal O&M is desired, a septic tank followed by a buried irrigation system should be chosen.

Based on personal experiments in his living lab, the Chairman is convinced that aerobic treatment should be avoided whenever possible. His main concern with aerobic treatment is its higher production of sludge and its dependency on an energy supply. Anaerobic treatment followed by a subsurface irrigation and seepage system is, as he sees it, the most appropriate solution for household wastewater treatment in developing countries.

Peak-hour pipe versus siphon. The even distribution of water from the septic tank onto the irrigation field was another technical issue considered. The Chairman ultimately installed a peak-hour pipe but had earlier also considered the use of a siphon.

For a *peak-hour pipe*, the outlet of the septic tank is slightly elevated so that the irrigation pipe is loaded with water during peak hours (e.g., in the morning). The system is extremely simple: when the water level in the tank is high, the overflow is let into the irrigation system. The drawback may be the lack of assurance of an even distribution in the garden. Some areas (the ones closest to the septic tank) might receive more water than other areas. This can to some extent be solved by using simple valves in a distribution box to periodically close off the pipes closest to the tank, to ensure distribution to pipes farther away.

A *siphon* is activated when the water in the septic tank reaches a certain high level and the pressure in the siphon will generate a flush effect, which distributes the water under pressure into the irrigation system, resulting in a more even distribution of water to the full irrigation area. The drawbacks are here the cost of the siphon and the risks, however small, of breakdown because a small crack in the siphon could allow air or dirt enter the siphon and the pressure that generates the flush effect will disappear. Also, because siphons are underground and usually difficult to access, it is a relatively complicated task to repair one that is damaged.

The Chairman found that the peak-hour pipe had the most simple and robust technology, which for years had supplied sufficient irrigation to all his plants, but also that the siphon, especially for larger systems, provided an interesting alternative.

4.3.1.3 General Reflections and Wider Considerations

More extensive development and installation of on-site wastewater irrigation systems calls for further investigation of a number of issues. There is a need for more easily accessible, locally relevant guidelines describing the design of the irrigation system and, in particular, the preferred selection of plants. The plants should have a high water uptake and transpiration rate so that the majority of water is vented into the air. In addition, the vegetation should be robust and resistant to shock loads of, for instance, sulphates and phosphates from clothes washing and chlorine from housecleaning. Ideally, the plants should also have a relatively low growth rate, which would keep the level of manual labor to a minimum.

For on-site irrigation systems to substitute for traditional simple seepage systems, there is a need to include the design criteria in the local building regulations and by-laws. Presently, these regulations typically only deal with the basic requirements for septic tanks and simple seepage, but design criteria for on-site discharge through drainage and irrigation systems should be developed and added.

5

Constructed-Wetland Wastewater Treatment at Baan Pru Teau: A Low-Cost Cluster Community System

⟫⟩ 5.1 Supporting a Cluster of Houses

The tsunami of 2004 struck hardest in the Phang Nga province north of Phuket, Thailand. Whole villages were wiped out, and shortly after the disaster housing became high on the priority list. In some areas villages were rebuilt on location, while in other cases new townships were built at new locations and families from different villages were relocated to these new townships. Baan Pru Teau is such a new township, consisting of five housing estates, namely, the Thai Red Cross village consisting of 80 houses; the Rotary village with 80 houses; the Bor Tec Tueng Foundation village with 96 houses; the Krung Sri Ayudthaya village with 112 houses; and the Pornthip/Ricky Martin village of 50 houses.

Relief aid is often provided for specific (high-visibility) purposes by specific grants from specific organizations, and some purposes are clearly more popular than others. Housing is high on such agendas, whereas infrastructure such as roads and solid waste and wastewater management is often low on the list. This often results in housing estates being quickly built without having the proper infrastructure. Two years after the tsunami, four of the five new villages still had not been provided with infrastructure—the houses were there and drinking water was provided but there were no roads, no drainage system, no solid waste collection system, and no treatment of wastewater. Some villages developed into slums within few months after completion. This situation had not changed as of mid-2009, five years after the tsunami.

The Thai Red Cross financed one of the new housing estates, including all necessary infrastructure except wastewater treatment. The construction of the new houses was undertaken by the Royal Thai Army and the wastewater management system was made possible by a grant from the Danish government; it was designed and constructed by a team of national and international consultants. The overall project was under the auspices of Her Royal Highness Princess Maha Chakri Sirindhorn through the Thai Red Cross Council.

Construction of the Thai Red Cross housing estate included installation of a primary wastewater treatment system (septic tanks for black wastewater) and a covered combined stormwater and wastewater collection system. Effluent from the septic tanks and all greywater was discharged into the established drainage system. The outlet of the drainage system was located at the front of the village, and the initial plan was to discharge it into a large freshwater reservoir in the center of the township. Unless a more advanced wastewater treatment system was provided, this plan would ruin the water quality in the freshwater reservoir for water supply to the villages and would create a potential public health risk.

The wastewater treatment component came late into the housing project, when the township was almost finished (including the construction of the central drainage system). The task was therefore to provide the village with a wastewater treatment system designed to alleviate the health and environmental issues, and furthermore to provide a demonstration project for the use of nature-based wastewater treatment technology for small housing developments and poorer communities. The choice of technology, made by the design team consisting of international and national government wastewater specialists, was horizontal subsurface-flow constructed wetlands.

Because the wastewater treatment system initially could support only one of the five housing estates in Baan Pru Teau township (all of which were located around a lake), a holistic approach linking the treatment facility into a larger management scheme was developed. The Red Cross facility would provide the first of up to five possible constructed wetlands located in the same area, thus over time potentially providing wastewater treatment for the whole of the Baan Pru Teau township. A rough wastewater management outline for the whole township was prepared, proposing that the other housing estates should be linked via a pressure pipe to the constructed wetland area with a pumping station located at each estate. The constructed wetlands were purposely overdimensioned to accommodate one additional village, and an adjacent area was left vacant so the remaining villages could eventually be included. When the big picture was in place, the design and construction of the wetland was initiated (Fig. 5-1).

The constructed wetland project was composed of three main elements: installation of horizontal subsurface-flow constructed wetlands; landscaping; and installation of a pumping station and an odor reduction system.

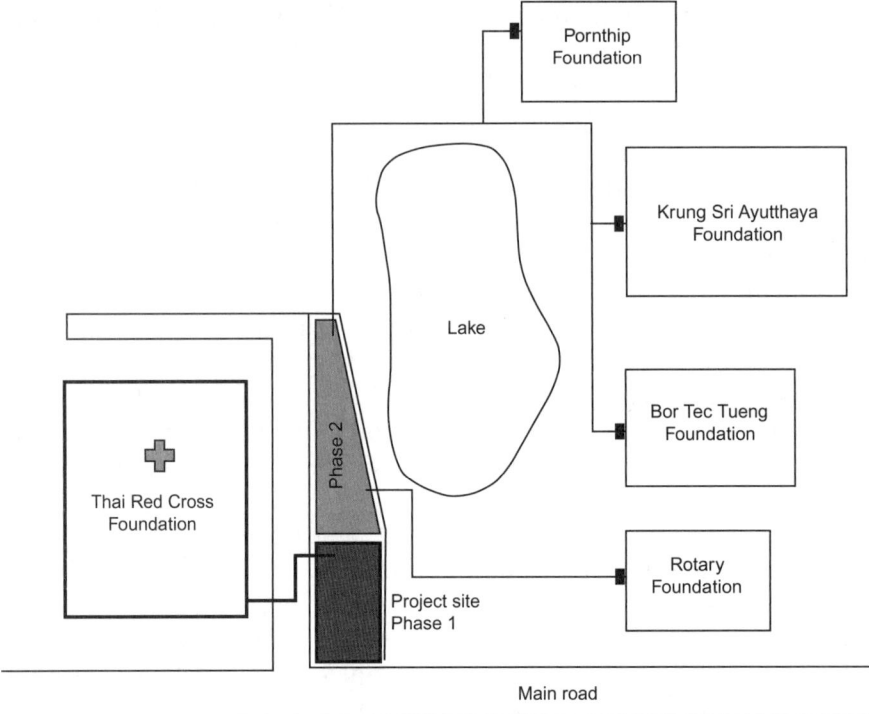

Figure 5-1. Site layout, including possible future Phase 2 extension.

A priority was that the system had to be simple and easy to operate and maintain. A number of basics were known: the wastewater would be characterized by low-BOD, often-diluted, mainly greywater; constructed wetland technology could effectively treat this kind of wastewater without requiring skilled technicians; high treatment performance could be achieved at much lower construction costs compared to conventional energy-intensive systems; and local affordability and skills were a problem in the tsunami-hit fishing village. Accordingly, energy-intensive and mechanical treatment systems would not be appropriate. The horizontal subsurface-flow constructed wetland system, planted with local emergent plants, was therefore selected as an appropriate wastewater treatment technology for Baan Pru Teau township.

The horizontal subsurface-flow constructed wetland was designed to treat wastewater from the 80 units of Thai Red Cross housing estate, and to be fed by a pumping station with an average flow rate of approximately 40 m^3 per day. The system consists of three units of horizontal subsurface-flow constructed wetland units in series, covering a rectangular area of about 600 m^2.

The wetland is located next to the main access road to the township. The first area proposed for the location was rejected by the governor, who considered it to be too centrally located. The design team therefore knew (not only from him, but also from many other projects) that two issues normally create the biggest resistance: How will the treatment system look, and how much will it smell? The focus was therefore on making the treatment plant as pleasant-looking as possible and counteracting the odor nuisance. Landscaping and odor control became key issues.

The design was finalized, a three-dimensional model was made for the governor, the approval was received, and after several discussions with the head of village (mainly on how he and the villagers could also get something out of the project—how win-win situations could be created, for example, by hiring the head of the village as a construction supervisor and by requiring the construction company that won the tender to hire and use workers from the village), construction was ready to start (Fig. 5-2).

The wastewater outlet had already been constructed in a low-lying ditch directly at the roadside. Because the land area allocated by the governor lay slightly above the outlet, final grading precluded a fully gravity-based collection and treatment system; there was no choice but to pump the wastewater once. A collection pipe was laid from the final manhole to the pumping station.

The pumping station was located at the top of the treatment area. The wastewater from the final manhole within the housing estate flowed by gravity to the pumping station. Aboveground, a pumping house containing a hoist, control panel, and odor reduction box was constructed. Bar screens, sand traps, the sump pump, and other pumps were placed belowground.

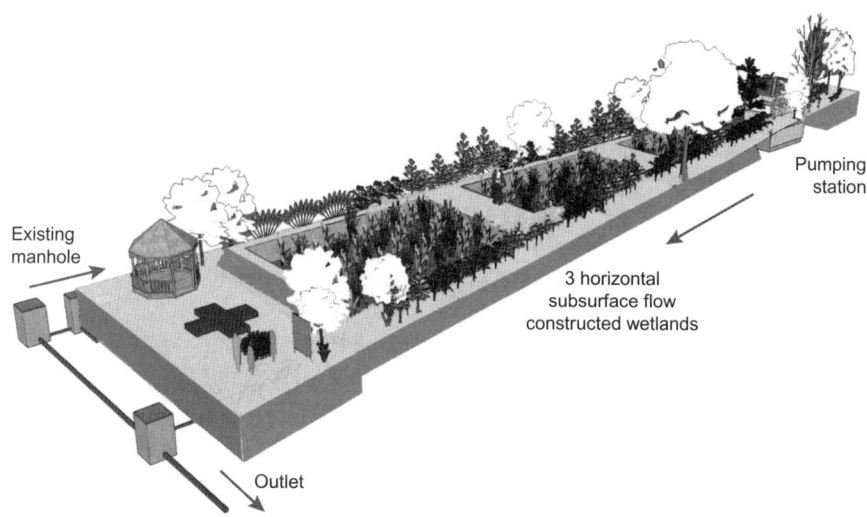

Figure 5-2. Computer model of constructed wetland at Baan Pru Teau.

From the pumping station the wastewater was lifted into the wetland systems, where the wastewater was distributed to the gravel filter by a perforated pipe. The rectangular wetland units each had a dimension of approximately 5 m × 10 m, each with a depth of 0.6 m and slightly elevated above the existing grade.

The outlet system of each wetland unit consisted of perforated pipes that discharged the effluent from the wetland units to the next unit's perforated inlet pipe, which had adjustable outlets to maintain and control the water level in the previous wetland unit. The adjustable outlets were able to regulate the water level between 5 cm above the surface of the bed to the bottom of the bed, to allow the complete emptying of the wetland cell as well as flooding it by 5 cm of water. To prevent groundwater infiltration, high-density polyethylene lining was used as a membrane in the constructed wetland units. On top of the gravel filter, the three wetland beds were vegetated by *Canna* (lily) spp. and *Heliconia* ("False Bird of Paradise") spp.

The open areas surrounding the wetland cells were covered with a 5-cm-deep soil/sand mix planted with Malaysian grass. For use as a rest area and to promote recreational activities, a Thai-style pavilion was placed at the entrance of the wetland site and benches were placed along the edges of the wetland cells. To raise public awareness, an information board including overview maps and descriptions of the wetland system were installed at the bench area nearest the main road. Outdoor ground lighting was placed around the site. All in all, it all looks much more like a park than a wastewater treatment plant (Fig. 5-3).

Even though construction began in heavy rain, it was finalized within two months. The governor, the design and construction team, and the head of the village were happy and satisfied with the result. Basic wastewater infrastructure had been provided for the tsunami relief situation—something most aid organizations normally do not want to touch because it is not very prestigious or is too complicated, with too many regulations and stakeholders needed to approve implementation.

))) 5.2 Reflections on Appropriateness and Sustainability

After the work was completed, an O&M manual was prepared and trained to and a large ceremony with more than 400 villagers was held, and the facility was handed over to the head of the village and the local village committee, with some anxiety. The anxiety stemmed from the fact that a treatment facility on this location, in this context, is up against quite a number of constraints, and only the future will measure the success of this facility.

Figure 5-3. The constructed wetland after 2 years of operation.

Reflecting on appropriateness and sustainability, the wastewater management system at Baan Pru Teau achieves only two out of the six "smilies" in Fig. 5-4. There was no strongly felt local need; there were no immediate public health problems this system would alleviate; and this was a weak, newly established community with high levels of poverty, unemployment, and uncertain land titles. Even though these issues were known from the start and had been addressed in the design of the wastewater management system, they created (and continue to create) uncertainties about each of the six elements in the wastewater management system, as follows.

5.2.1 The Collection System

An on-site system was not an option for the design team because the team was tasked with treating the water at a specific, already established outlet. Pragmatically, the team tried to make the best of the given situation, but the collection system was never an active parameter in the design. And maybe luckily so. The team would, if given the option and based on their experience and local observations, probably have proposed and installed an on-site soak-away system. However, at about the same time a neighboring village installed simple,

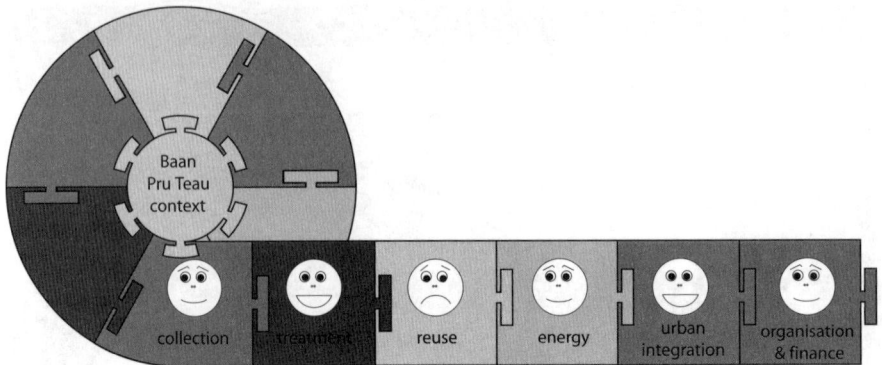

Figure 5-4. Contextual appropriateness scoring of the six elements of the wastewater management system at Baan Pru Teau.

Smile, contextually appropriate; no expression, somewhat appropriate; frown, not appropriate.

traditional seepage septic tanks, which from Day 1 did not function due to the local impermeable soil. This resulted in serious problems with flooded septic tanks and smelly wastewater on the ground surface during rains (ironically, seepage tanks 500 m away worked perfectly because that soil was different). Normally, it is enough to look around the neighborhood and ask questions, but in this case an actual soil analysis would have been required to design a suitable soak-away system on this particular site.

5.2.2 The Treatment System

The villagers in Baan Pru Teau are predominantly experienced fishermen, boat builders, and manual laborers. The level of education and technical wastewater management expertise within the village is limited, and it was therefore necessary to build a facility that did not need much O&M skill. Accordingly, the treatment method was based solely on natural processes inside a gravel filter and, once in the wetland cells, was designed to work by gravity only. The O&M activities required to run the facility consist mainly of pump operation, adjustment of water levels in the wetland cells, and removal of excessive plant growth. Local manual labor could accomplish this in weekly or monthly intervals.

5.2.3 Energy

Ideally, the design should not have been based on a pumping station. Pumping certainly is the weakest link of the design for this location. If the treatment system fails, it fails because the pumps are switched off or broken and not repaired. The need for a pumping station introduces an unfortunate need for power supply and energy consumption for the operation of the treatment facility. Had the planners of the village infrastructure included wastewater treatment from the beginning, the demand for energy could have been reduced or even eliminated. This unwanted pumping station might eventually render the facility nonfunctional. However—and this is the advantage of constructed wetland systems—if such a facility is abandoned, it can relatively easily become functional again if political and financial priorities change. Conversely, when mechanical systems are abandoned they become useless and almost impossible to reactivate because infrastructure and equipment deteriorate, break down, or get stolen.

5.2.4 Re-Use

The assigned task was to collect wastewater at a specific outlet and build a wastewater treatment plant for the township, making treatment the primary issue. Re-use did not become an immediate priority. Resolving the social, economic, and legal issues of the site, getting the basic project in place, and reducing the level of complexity seemed to dissolve the importance of an integrated re-use system. With so many other things that could go wrong, why add this element of complexity? Looking back on the design process, more persistent consideration should have been given to the re-use and re-entry issue. In poorer communities it is likely that treated wastewater can be utilized for irrigation, gardening, ponds for fish farming, or the like. Any integration of wastewater re-use and income generation would contribute to increased chances of sustainability and should be considered and pursued to its fullest.

5.2.5 Urban Integration

When asked about their preferences regarding a wastewater treatment facility, the inhabitants and the village chief voiced concerns about location, appearance, and odor. To gain acceptance for the project, these issues were high on the agenda during planning and design. This highlights the importance of thoughtful urban integration to secure community acceptance and participation, both of which are imperative parameters for success.

The location of the facility—only few meters from private houses—engendered the idea of constructing an area with the appearance of a green space or a public park. Plants were selected primarily to beautify the area. Flowering plants were chosen for the wetland cells and a variety of trees and shrubs were planted along the edges of the treatment facility. Having acquired a wastewater treatment plant, the township has also been blessed with a communal green space of high quality. In a traditional engineering approach, most or all attention is put on technical treatment optimization; however, the Baan Pru Teau example shows that multiple parameters contribute to the appropriateness and sustainability of the system.

Foul odor from the collection system, the pumping station, and the treatment facility was minimized by installing septic tanks at each household, by installing an odor control box inside the pump house, and by using subsurface treatment in gravel filters. The odor box in the pumping station was equipped with a ventilator fan that creates a slight vacuum in the sump and the collection pipe. The air drawn from the system is released through an odor filter consisting of pieces of charcoal, wood chips, and straw. Odor from the wetland cells is reduced by the subsurface flow, where the wastewater surface is kept some 10 cm below the surface of the wetland. Only at the inlet section at each cell is the wastewater on the surface.

5.2.6 Organization and Finance

Wastewater treatment plants are normally taken over and run by the local authorities. However, as this project progressed it became apparent that the treatment plant serving the newly erected village had no land title because it was located in an area outside of both municipal and regional responsibility, on state land administrated by the governor, which locally is more or less an administrative no-man's land. This challenged the organizational and financial setup in terms of operating and maintaining the system. For instance, without a land title, the electricity company did not have an address to send the bills to, so an agreement had to be made with the village person in charge of power connections.

Because of this special situation, a 3-year O&M contract was included in the construction contract to ensure that the community could keep the facility in operation, pay the water bills, keep the park attractive, and so forth for at least 3 years. The village head was contracted to secure efficient O&M, and well into the third year he has showed a continued high level of commitment and responsibility through effective O&M of the facility. Nevertheless, the destiny of the system when the 3 years of O&M support is over is yet to be seen. A permanent solution could not be established during the initial planning, involvement, and construction in Baan Pru Teau, and this can prove fatal. Should no long-term decision on O&M be made, no central funding be allocated, or no fee collection system be initiated, the facility could face serious operational problems.

In terms of organization and finance, the design team tried their best. The head of the village was directly and actively involved during both construction and postconstruction; the local work force was used; and 3 years of O&M donor support was secured (Fig. 5-5). Still, this does not solve the basic

Constructed-Wetland Wastewater Treatment at Baan Pru Teau 109

Figure 5-5. (*Clockwise from lower left*) Speech by the head of the village, Mr. Bornsong Chaysawaay; the construction team from Envire Scan Co. Ltd. (Thailand) led by Dr. Chatdanai Jiradecha (*first from right*); the operations manager getting training; and the community invited to an information and dining event when the system was put into operation.

problems associated with this site: the fact that the local people have experienced so many constraints in their lives; the absence of a clear administrative body for the ongoing operation, management, and financing of the facility; problems with land titles; the high unemployment rate; and some of the residents possibly returning to the sea shore in the coming years. This is the reality of poverty and this is what the facility, and the design team, were up against.

Now into its third year of operation, the facility is still well kept, effectively maintained and operated, and functions as a showcase for the local community in charge of it. Perhaps in this case ours is too pessimistic an assessment. Sustainability can be and is grounded in many things. Sufficient budgets, skilled staff, and available spare parts are the typical ones, but perhaps in this case the fact that the constructed wetlands have developed into a small showcase can prove to be the factor that secures sustainability (since its inauguration it has had two or three official groups visiting the facility almost every month). Who knows?

5.3 Smart Technologies at Baan Pru Teau

The two most interesting technologies utilized in the wastewater management system at Baan Pru Teau are the horizontal subsurface-flow constructed wetland, and urban integration and the importance of landscaping (Fig. 5-6). The first technology is discussed at length in Chapter 8 and the landscaping is discussed below as well as, in more detail, in Chapter 6.

5.3.1 Landscaping as a Design Parameter

Constructed-wetland wastewater facilitates can upgrade local environment both aesthetically and socially through careful landscaping and beautification. At the Red Cross village in Baan Pru Teau, the wetland was located and shaped to form a welcoming garden at the entrance to the community. Turning into the village from the main road you are welcomed by a lush flowering park flanking the access road on the right side, just opposite the first row of houses. This green gesture tells the story of a housing estate symbolized with a pleasant, eye-catching landmark, the Red Cross Garden. Blooming Canna lilies, Travellers palms, heliconias, lush green grass, and decorative shade trees lining and encapsulating the area create an inviting garden environment with full public access, and enhance the visual identity of the site.

The rehousing project itself is densely built in order to accommodate as many tsunami-stricken families as possible within the available land. Houses are built using relatively cheap materials and are completed very quickly. Combined with high unemployment and the threat of poverty, this new village and the surrounding area are balancing on the razor's edge of developing into a slum. By adding and shaping the constructed wetland for wastewater treatment into a public park for the benefit of the villagers, an optimistic social and aesthetic agenda tries to counteract the possibly dim future of the township (Fig. 5-7).

The wetland is located on a marginal triangular land strip, compressed to optimize land use in the village for dwellings. Still, both functionality and treatment performance can be achieved without compromising public acces-

Figure 5-6. Landscaped constructed wetland for cluster wastewater treatment for a small remote village: a smart technology?

Smile, contextually appropriate; no expression, somewhat appropriate; frown, not appropriate.

Figure 5-7. Landscaping (Canna lilies) at Baan Pru Teau.

sibility. The Red Cross logo is integrated into the landscape design as a flower bed and symbolizes the linkage to the international aid organization that sponsored the township after the disaster. The logo symbolizes the identity of the village and distinguishes it from the surrounding communities. The symbolic garden can potentially help unite the community by adding a common element which they, as a group, can be proud of—shared social confidence.

5.3.2 Technical Considerations, Balances, and Choices

Because this was a demonstration project, much effort was invested in making the landscaping as attractive as possible. One could criticize landscaping for using budget money for plants, flowers, and shrubs rather than for improving treatment efficiency, but the project showed that, in general, landscaping does not incur significant additional costs. Grass seed, some trees, and a few benches is all it takes. In addition, the choice of plants can reflect robustness and need minimal maintenance.

If it is suggested that a wastewater treatment plant be located close to a residential area, a common, almost automatic design solution is to fence off the entire facility. This is due to public health risks but also perhaps to a tradition developed at large, centralized systems with expensive equipment in danger of being stolen. Small-scale wastewater treatment facilities might not have the same need to be fenced off. Only a limited number of items can be stolen at Baan Pru Teau because all valuable equipment is locked inside the pumping station. Furthermore, the public health risk of a subsurface-flow constructed wetland is low.

If one wants a wastewater treatment plant to double as a public park or recreation area, barbed wire and fences are not the best way to invite visitors in. Promoting rather than restricting access to the Baan Pru Teau facility enhanced its usability and encouraged the local community to "adopt" the system. The openness also enabled more watchful eyes to prevent misuse, damage, and violation of the park environment.

5.3.3 General Reflections and Wider Considerations

On the day of the grand opening, the constructed wetland was a blooming flower park. However, also on that day several kilos of litter were collected from the site. Plastic bags, candy boxes, straws, and other waste had been thrown into the bushes and had piled up on the site within only one or two months. This did not portend long-lasting attractiveness for the beautified wastewater treatment plant in the fatigued environment. Three years later, the park is surprisingly well maintained. Against all odds, the locals have taken responsibility for keeping the area in good shape and the stated intention of creating a space that the community as a whole could be proud of seems to have come true. Maybe this is also a result of choosing trees and plants that are beautiful but need water and maintenance to keep their beauty. Perhaps by providing the *opportunity* for responsibility and local involvement, responsibility and local involvement ensue.

The introduction of wastewater treatment systems shaped as public parks or gardens introduces a humane and natural appeal to the local residents, contrary to the impression of alienating concrete structures. This opens up a completely new view on treatment systems and their potential integration in the urban environment. By adapting systems based on natural treatment processes, wastewater facilities can be changed from a somewhat negative, unpleasant, and introverted activity hidden behind walls, to an extroverted, inclusive, and even beautiful urban element that can develop into a catalyst for urban design and environmental consciousness among the public. The potential of landscaping and integration in the urban environment is enormous and the exploration of possible ways to integrate the design of constructed wetlands into gardens, parks, playgrounds, and ponds creates a whole new field of study for gardeners, architects, landscapers, and urban designers.

In addition, by adopting such an approach in urban planning, the traditional monofunctionality of conventional wastewater treatment systems can be left behind. The design potentials of constructed wetlands allow for a layering of multiple functions, such as aesthetic upgrading, public parks, identifying landmarks for housing estates, local nurseries, or small, nonintrusive treatment systems implemented on marginal urban spaces such as roadsides, back yards, parking lots, or urban wastelands encapsulated between infrastructures. These marginal spaces can be activated and utilized for improvement of the urban environment through landscaped wastewater treatment systems, and thus be reintegrated more actively in the city.

6

Wastewater Management Design at Koh Phi Phi: A Recovery-Based, Closed-Loop System

))) 6.1 The Flower and the Butterfly

From the hammock on the hillside you watch the sun rise above the silent bay. The chatting and laughter from a group of young, blonde Scandinavians has caught your ear. The group has just arrived on the beach to have their morning swim and refresh themselves after last night's Christmas dinner.

The turquoise water, the swaying coconut palms, and the tranquility of the twin bays embrace you. Paradise truly exists (Fig. 6-1).

6.1.1 The Wave

Simultaneously, some thousand kilometers from this bountiful island, two tectonic plates rub shoulders deep below the Indian Ocean. A powerful earthquake about to evict 300,000 human beings in all affected countries has emerged. The movement of the sea bottom creates a tidal wave, traveling at 800 km per hour toward Thailand and Indonesia to the east and Sri Lanka, India, and the African continent to the west. The 2004 tsunami disaster of the Indian Ocean is about to become a reality.

You sway peacefully in your hammock. The shallow water and the low tide have extended unusually far from the beach this morning. The Scandinavians have finished their swim and are now hunting seashells on the dry sea bed. Several tourists join the hunt and enjoy this spectacular sight. But just under the winds whispering in the palm leaves, you sense anxious talk and faces expressing concern. The local islanders seem uneasy.

Figure 6-1. The sand strip of Koh Phi Phi prior to the December 2004 tsunami.

At the speed of an airplane, a huge wave bends around the western mountain and strikes the island first from the south and then from the north, eradicating almost everything in its path. Soon a second and even more devastating wave hits the island. Within minutes, most man-made structures on the densely built-up strip of land are washed away. Bungalows, shops, and restaurants are in ruins. As many as 600 human lives are lost on the island.

A few coconut trees sway on the barren land strip (Fig. 6-2). Tranquility again embraces the island.

6.1.2 Background

Phi Phi Island is located in the archipelago of the Andaman Sea on the west coast of southern Thailand. Two hours by boat from Phuket, you arrive on an island in the heart of a national marine park. Two mountainous rock formations completely covered by rain forest rise vertically hundreds of meters above sea level. They are connected by a narrow sand dune about 1.5 km long and 200 m wide, creating two U-shaped lagoons and making the contour of the island resemble the shape of a butterfly. Collectively, the two formations on either side and the sand strip are known as Koh Phi Phi Don, translated as "the island of the 'phi phi' trees and the sand dune." The mountainous parts of the island are preserved as a part of a national park, whereas the sand strip has historically been divided among and developed by a number of private landowners. As such, this commercial strip of land is the pearl of the marine park.

During the first decades of Koh Phi Phi's development into a tourist destination, all dwellings on the island were equipped with seepage wells and

Figure 6-2. The sand strip of Koh Phi Phi after the December 2004 tsunami.

all wastewater was treated and discharged within the plot. A common collection system or a central treatment plant was not needed. As time went by, the increased densification of the sand strip, the limited availability of land, and the fact that the seepage wells had to be moved and rebuilt every 5 to 10 years due to geological conditions and a high groundwater table, the on-site wastewater management system reached a critical mass. It seemed reasonable to implement a centralized municipal collection and treatment system that served the densest built-up areas of the island.

In the early 1990s a group of government authorities, consultants, and contractors arrived on the island. They funded and built a collection and treatment system—and left the island. What the group did not do was consider parameters such as local needs, impact on neighbors, handing over the system to the local authorities, training local staff, operation and maintenance, or whether the installed facility was suitable or sustainable to serve the island. History proved that it was not.

The wastewater treatment plant was never put into operation (Fig. 6-3). For a decade the municipality had two large ponds, fenced behind barbed wire, that did not receive any wastewater. Why? The most evident reason was that parts of the gravity-based collection system had been constructed with a negative slope. Most of the wastewater simply accumulated in the lowest point of the sewer line somewhere in town and never reached the treatment plant. The wastewater decomposed in the pipes under the streets, causing strong

Figure 6-3. The sand strip of Koh Phi Phi after the December 2004 tsunami with the previous pond treatment plant in the lower left corner.

odor problems and, during rains, flooding into the streets. Reality was the opposite of a tropical paradise. This, of course, was an intolerable condition for the residents and the tourism industry, and the municipality reacted by implementing a by-law *prohibiting* people from discharging wastewater to the collection system. It is an awkward situation to have a collection system where people were fined if they connected to it! Without flow through the system, the wastewater collection system slowly filled with sand. The entire investment and the planned environmental and tourism-based economic benefits of the sanitary system were lost, mostly because of poor planning and the absence of local involvement.

6.1.2.1 Planning Process
When communal systems fail, people have to manage and solve the problems themselves. When the municipal collection system on Koh Phi Phi failed and the municipality banned people from connecting to the system, a local hotel/land owner, Ms. Witchuda Jantharo, took over and implemented a second independent, parallel privately owned and operated collection and treatment

system on her own part of the island, which was about half of the island! No municipal involvement, no governmental budgets, no international consultants—just an identified need for sanitation mixed with private entrepreneurship. And the best part: wastewater was collected and the system worked! An advanced private wastewater treatment at the end of the collection system was under construction when the tsunami hit the island.

6.1.2.2 Rehabilitating Paradise

The tsunami left Koh Phi Phi devastated, with its businesses and infrastructures destroyed. Utility networks, roads, water supply, electrical power supply, and the wastewater drainage system were all in ruins. Deeply affected by the impact and scale of the natural disaster, the international community showed human generosity and great willingness to donate. One of the projects financed through such donations, a grant from the Danish government, was the rehabilitation of the wastewater management system at Koh Phi Phi.

After the tsunami, the public wastewater collection system remained partly intact, whereas Ms. Jantharo's private treatment system was completely devastated. She was about to rehabilitate her treatment system when the idea of linking up with the municipal rehabilitation project emerged as a win-win situation for all involved parties. Ms. Jantharo was saved the expenses of building a new treatment plant and the municipality was given access to her well-functioning collection system, which would ensure adequate loadings at the new municipal treatment plant.

A handshake agreement enabled Ms. Jantharo to provide land for a pumping station and allow other landowners to connect to the established collection system; the municipality to provide land for the treatment plant and collection pipes to other landowners; and the designers and donors of the new wastewater management system to begin their work. Two local public hearings and meetings with central and provincial authorities provided the go-ahead for the project. Having the local administration, the local entrepreneur, and other landowners joining hands created mutual technological, environmental, and economic benefits.

The mayor, Mr. Phankhum Kittitarakhun, who had learned the lessons of the previous wastewater management system on Koh Phi Phi, demanded that a 5-year contract covering O&M of the entire system be included in the design and construction contract with the international donor before any civil works could commence. It turned out that Mr. Kittitarakhun had done independent study trips around the region to study best practices of wastewater management, and personally believed that constructed wetlands were the most suitable solution for treatment of wastewater in tropical areas. This came to influence the choice of technology in this project.

To ensure that the project met the actual needs of the island and not just mutual economic interests conspired by politicians, contractors, and government agencies, local residents and stakeholders were involved in an early stage in the planning process (Fig. 6-4). The community leader on the island, Mr. Sommai, was hired as the mediator between the locals, the mayor, and the design team, which consisted of international and national wastewater specialists. The mayor allied with an independent local technical expert to ensure integration of public and municipal interests in the project and to evaluate the appropriateness of suggestions from the design team and the contractor. Public hearings and a vote on system requirements and the physical layout were held. Business leaders were involved in terms of interests, finance, and motivation.

A basic principle was to pursue mutual benefits for all stakeholders. The specific need for a wastewater management system became the key component that at the same time could facilitate the realization of an integrated urban design and environmental management plan for the island.

6.1.2.3 Environmental and Symbolic Integration, Multifunctionality

The treatment plant was to be built in a prominent location that all visitors pass while strolling the island, which heightened the need for an aesthetically attractive facility that would make a positive impression of Koh Phi Phi as a bountiful island. The facility was designed to resemble a butterfly sitting on a flower—a symbolic reference to the butterfly-shaped contour of Koh Phi Phi. The relationship between the flower as a living organism and the butterfly as a carrier of pollen symbolizes the new beginning, the growth and bloom of the flower, the community, and the island in the aftermath of the tsunami (Figs. 6-5 and 6-6).

Given the limited amount of land available at Koh Phi Phi, as well as to ensure that the project will give good value for the money, the wastewater treatment facility has a multifunctional design that optimizes land use and facilitates spin-offs with mutual benefits for the municipality, landowners, residents, and tourists on the island. Not only does the system treat the waste-

Figure 6-4. (*Left to right*) The community leader, Mr. Sommai; the mayor, Mr. Phankhum Kittitarakhun; the primary landowner, Ms. Witchuda Jantharo.

120 *Sustainable Wastewater Management in Developing Countries*

Figure 6-5. The wastewater collection system with secondary pipes (*black*), mains (*bold black*), pressure pipe (*white*), re-use water pipe (*dashed white*), pumping station (*round dot*), and inlet structure (*square dot*).

water on the island, thus securing public health and a clean physical environment, it also functions as a public park with walk paths, benches, and a pavilion. Everyone can enter the park and enjoy the blooming flowers as well as become informed about the processes in the system, thus learning about water treatment and the importance of sustainable environmental management. In addition, space is provided for *sepak takraw* ("kick volleyball") and

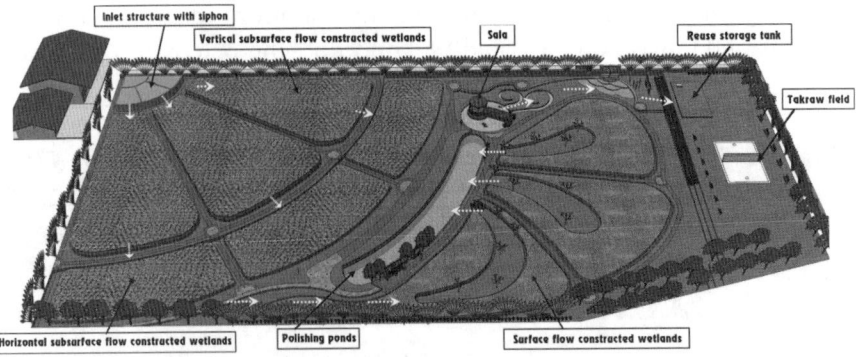

Figure 6-6. Design drawing of the constructed wetland system.

other leisure activities. Given the park's layout and multifunctionality, the term "wastewater treatment plant" could be replaced by the much more accurate "water reclamation park."

Because Koh Phi Phi is a small island located in a national park and is highly dependent on the tourism industry, the maintenance of crystal-clear water at the beaches is essential. Therefore, it would be detrimental to discharge wastewater onto or near the beaches. Rather, the wastewater had to be regarded a resource to be re-used for irrigation purposes. This would be a benefit for the tourist-related businesses relying on green lawns, blooming flowers, and lush trees, especially because the reclaimed water could be sold at a fraction of the cost of tap water. This would benefit the environment and the island community as a whole by counteracting the water scarcity problem.

6.1.3 Design: An Integrated Cluster Wastewater Management System

The final design included all components of wastewater management: wastewater collection, treatment, urban integration, re-use, energy, and organization and financing of the O&M (Fig. 6-7). The project was built and today operates as follows.

Almost all of the wastewater from washing, bathing, and cooking (the greywater) is discharged to a closed-loop collection system, separate from rainwater. Some, however, is discharged to semi-open drains before being connected to the closed-loop, small-pipe system. Most hotels and restaurants have installed grease traps within each compound to prevent oil and greasy wastes from clogging the collection system as well as the municipal treatment facility. However, some of them—especially those coming into operation late in the project or after project completion—have still (as of mid-2009) not installed oil and grease traps. The wastewater from toilets is collected and pretreated in local septic tanks. The effluent from the septic tanks is discharged either to the closed-loop, small-pipe system or to the semi-open drains that collect the greywater from households.

In the areas where the closed-loop, separate collection system is installed, this system receives and transports only domestic wastewater. Stormwater run-off is managed with an independent drainage system. This way, the risk of sanitary wastewater reaching the streets during heavy showers is minimized and sand is prevented from entering and blocking the pipes, which ensures functionality and minimizes the maintenance costs of the system. Moreover, the wastewater is not subjected to dilution, which ensures a relatively constant level of treatable constituents and optimizes the design criteria for the wastewater treatment facility.

Figure 6-7. The constructed wetlands on Koh Phi Phi 2 months after completion, November 2006.

Because there is no law forcing households to connect to public wastewater collection systems, the team decided that construction of the collection area of the project would include connection of households located there. This meant that the construction contract included service pipes and actual connection taps for households that needed and wanted to connect. The contract also included a number of septic tanks and oil and grease traps for households, restaurants, and hotels that were willing to install them to enhance the efficiency of the system.

All wastewater is collected by gravity flow and, because oils, greases, and solids are less likely to reach the local treatment units, the system is unlikely to clog. The gravity-flow system collects all domestic wastewater to a single location in the central part of town, from where it is pumped to the treatment facility (Fig. 6-8). To prevent odor problems, the pumping station is equipped with an odor control unit.

Technically, the treatment facility can treat up to 400 m^3 of wastewater every day, which is treated by a mix-and-match of four different treatment technologies. As the wastewater flows through the treatment facility, it passes through a vertical subsurface-flow constructed wetland, a horizontal subsurface-flow constructed wetland, a free-water surface-flow constructed wetland, and a pond. The vertical gravel filter treats the wastewater by removing 80% to 90% of the organics, nutrients, and pathogens from the wastewater. From the vertical-flow wetlands, the water flows into a series of, first, hori-

Figure 6-8. Images from the process. Three-dimensional illustration used for public hearings and decision makers (*top left*); the mayor's technical supervisor, Mr. Pisit Srivilairit, and the contractor, Mr. Niras Limprayoonyong of Mahaporn Co. Ltd. (*bottom left*); and two international consultants, Dr. Hans Brix (*left*) and Mr. Carsten H. Laugesen doing on-site inspection (*bottom right*).

zontal subsurface-flow wetlands, and then surface-flow constructed wetlands, and finally into a polishing pond before it is discharged to the re-use reservoir. These components are all very simple and require almost no maintenance.

To reduce energy consumption in the collection system and the water reclamation park, the pumping station in the town center is equipped with solar panels to provide electricity to operate the pumps. The solar power station was designed to operate one pump for 6 hours each day. The treatment facility intake is dosed using a siphon instead of an electrical pump, which reduces the system's energy consumption because the siphon works entirely on hydraulic principles and without a power supply.

The reclaimed water and flowers from the wetland units are sold to private landowners and the revenue is designated to cover some of the O&M costs. This supports the long-term functionality and appearance of the wastewater treatment system. By generating income from its operation, the facility has a better chance of becoming a sustainable component in the island's infrastructure.

A local contractor, Mr. Niras Limprayoonyong of Mahaporn Co. Ltd., built the project, and he hired members of the local community committee as advisors and middle managers. This imparted knowledge of the local context, local suppliers and local pricing, and loyalty and mutual responsibility. This also developed local expertise on the functionality of the system, the facilities, and individual installations—all important issues for the O&M of the system in the years to come.

After construction was completed, an environmental fund managed by a local environmental committee was established to secure the ongoing O&M of the facility. The fund would receive money from the 5-year O&M contract signed with the donor, from the sale of flowers and re-used wastewater, and from wastewater connection fees for new businesses and hotels that connect to the system. All this income is to be spent on O&M (staff, power supply, etc.) and promotional activities. The committee consists of the mayor and members of the local community.

The project has become a showpiece of integrated cluster wastewater management. It not only meets the demand for urgent rehabilitation of the wastewater infrastructure on the island, but also addresses the island's specific limitations on water and energy supply as well as the need for cyclic management and self-reliance on an island with limited natural resources. The project represents the essence of what we consider appropriate and sustainable cluster wastewater management, as the integrated system on Koh Phi Phi has attempted to apply these principles in practice.

))) 6.2 Reflections on Appropriateness and Sustainability

The six-element checklist indicates, first, whether all elements in the wastewater management system have been dealt with, and second, roughly to what extent each element is considered to fit the local setting (Fig. 6-9). As can be seen, the Koh Phi Phi system does contain all six elements in the management system and scores high on four of them.

6.2.1 The Existence of Real Needs—It Makes Sense!

The reasons why wastewater at Koh Phi Phi should be collected and treated are obvious to almost all landowners, residents, and tourists on the island. The case certainly would pass the "does it make sense" test. Without wastewater treatment, the tourist industry would quickly and visibly be negatively

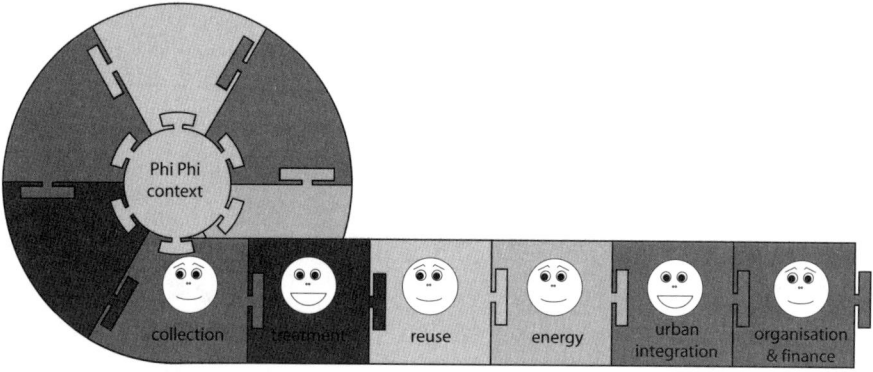

Figure 6-9. Contextual appropriateness scoring of the six elements of the wastewater management system at Koh Phi Phi.

Smile, contextually appropriate; no expression, somewhat appropriate; frown, not appropriate.

affected. Wastewater discharged directly to the beaches, wastewater overflowing from seepage tanks into the streets and onto the ground, and toilets that cannot flush are not many tourists' image of a small tropical paradise island. Without a centralized collection system, many of the households will have problems getting rid of their wastewater because local groundwater levels are periodically too high, and land parcels are too small for full seepage systems to be relocated (which is the traditional way of emptying septic tanks: relocate it 2 m away!). This means most of the influential landowners and many of the smaller households have a direct and objective interest in the island having a well-functioning wastewater management system in place.

Other contextual needs on Koh Phi Phi are typical island issues of scarcity. The first is water scarcity. Water supply depends on seasonal levels of precipitation, and the island often experiences periods of insufficient water supply. Most of the water supply is private and very expensive, so water becoming wastewater is a luxury. A wastewater management system that could return some of this water, at a lower cost, would be highly desirable.

Next is land scarcity. With only the middle low-lying strip of approximately 30 ha (0.3 km^2) being available for private or communal ownership (the hills are all protected national parks), and with more than 1 million tourists visiting the island yearly, land is scarce and valuable. Providing a 6,000-m^2 central wastewater treatment facility on the only tract of municipal land on the island, thereby reducing land requirements for private wastewater treatment, is therefore appreciated by the private landowners.

Finally, there is energy scarcity. The island is located about 40 km from the mainland and all electricity is currently supplied by private diesel generators, which are expensive, noisy, and annoying to residents and visitors. Providing a wastewater management system with low energy needs at least does not aggravate the energy supply situation.

6.2.2 Taking Local Issues and Stories Seriously

At the first meetings between the mayor and the design team, the mayor emphasized three requirements that he and the local citizens considered most important:
- The new system should not smell bad.
- It should look beautiful.
- It should be easy and cheap to operate and maintain.

All these requirements were closely linked to previous and existing issues and stories on the island. For example, residents and hotel owners had rejected the previous wastewater collection and stabilization pond system, mainly due to its odor problems. As a consequence, the system was ultimately shut down and the ponds turned into unsightly stormwater-filled ponds. This experience had taught local residents, landowners, and the municipality that a wastewater treatment plant located adjacent to dwellings can be a nuisance to its neighbors.

Given this history, no one (and especially not the mayor) wanted to risk reintroducing problems connected to stagnant, smelly wastewater if the wastewater treatment facility was to be rehabilitated. This created an unequivocal demand for an odor control system in all components of the system—in the collection system, at the pumping station, and at the constructed wetlands. If the odor control system were to fail, public support would fail and the project as a whole, and the mayor in particular, would come under pressure.

Beautification and easy and cheap operation were the two other important local issues, which from a design point of view simultaneously become constraints and opportunities. Both issues are dealt with in the following sections.

Apart from these three paramount issues, a long list of other local constraints and opportunities had to be taken into account, such as the existing infrastructure, the mayor's preferences, the available budget (a low-budget donor project), the location, and the landscape.

6.2.3 Creating as Many Win-Win Situations as Possible

The winner in wastewater treatment is often thought of as the environment, but this is rarely sufficient. Win-win situations have to be sought, found, and created for as many stakeholders as possible. Even though motivations are often personal and hidden, and therefore not so easy to predict or get right, some of the positive motivations—win-win situations—established on Koh Phi Phi include:

- *The mayor.* Who knows what a mayor gets out of such a project? But one thing is certain: If Mr. Kittitarakhun had not gotten anything, the project would not have happened and this chapter would not have been written. He was the key person in this case, and this is probably more or less the truth of island municipalities. Reputation, personal motivations, an honest wish to get things done better—the motivations might be many and interlinked. Interestingly, long before the tsunami disaster, Mr. Kittitarakhun had on his own investigated the possibilities for constructed wetlands on the island. This project provided him with a chance to capitalize on this interest and to create a showcase, a first, in Thailand.
- *The contractor.* Besides the opportunity to make some money, this project offered Mahaporn Co. Ltd. a strategic opportunity to establish itself in a possible new market. Mr. Limprayoonyong had a previous close connection to the mayor and considerable working experience on the island. In fact, before the construction was finalized on this project, he had secured two contracts with another municipality and a large industrial estate to design and build constructed wetlands for wastewater treatment.
- *The key hotel and landowner.* Because of this successful collaborative project, Ms. Jantharo did not have to rehabilitate her damaged advanced wastewater treatment plant and could avoid this considerable cost by linking to the new municipal treatment plant.
- *The local community leader.* Mr. Sommai was hired by the contractor as construction supervisor and was later put in charge of O&M of the system.
- *The local residents.* They were relieved of the increasingly difficult problem of finding new locations for their seepage systems.
- *The international and local consultants.* We were given the chance to work together with a group of honest, hardworking people (the mayor, the contractor, the local community leader) who not only wanted the project, but also wanted to implement it to a high standard. We were able to design, innovate, and bring to full implementation an interesting project within our professional field of interest—and hopefully the design was something to be proud of. Such a chance does not happen every day in the wastewater management sector.
- *The others.* The governor, the Ministry of Natural Resources and Environment, the Wastewater Management Authority, the National and Provincial Public Works Departments, the Danish government

as the donor, the participating universities and consultants, the sub-suppliers, the gravel freight company—the players involved in the design, approvals, tendering, contracting, and implementation were diverse and numerous. A key factor for success and sustainability, and probably the most important task for the project manager, was that at the end of the day all of these involved actors had a win-win feeling.

However, the other side of this coin is that the involved parties have something at stake—if the system fails, those involved will lose face, which is an extremely strong motivator. We will lose face if we have guaranteed to our peers, constituents, supervisors, or bosses that the project will work. The issue here was to develop as many interdependent relationships as possible: the contractor dependent on the mayor; the mayor accountable to the governor and his voters; the consultants dependent on their reputation in the international professional wastewater field; the community leader responsible to the powerful landowners on the island; and so on.

Exposure is important in this respect. A system that was built and failed in a small municipality in the middle of nowhere would attract little attention. But a system having such exceptional design and landscaping, which has been given exposure in newspapers and professional magazines, which weekly has groups of visitors from the Ministry, from abroad, from NGOs, and from technical experts from all over, would be more difficult to let fail. It could still fail, but this is less likely because so many people have something at stake and could lose face.

6.2.4 How Could This System Fail?

With the completion of the construction, a good foundation for efficient wastewater management on Koh Phi Phi is in place. The system is contextually designed; it is appropriate; it makes sense. But it must be stressed that it is a *design*—a design that has just been implemented and set in operation. If the system runs into serious O&M problems, or even fails, a number of possible reasons exist.

The system is up against a historically poor track record in Thailand, on islands, and in particular on Koh Phi Phi. When reflecting on sustainability, it might sometimes be good to think of the big picture. What is the system up against? This means that sustainability is not seen in the singularity of an individual project, but in the multiplicity and complexity of a historical, political, and national perspective.

Wastewater management systems in Thailand have a troubled history—almost all of them malfunction soon after implementation. On islands, such systems are up against what could be called the "island culture of small communities," with shifting strong and weak community leaders and organizations and high levels of corruption and infighting. Koh Phi Phi, in particular, has a long history of malfunctioning, inoperative, and unmaintained infrastructure projects—wastewater as well as energy supply, water supply, and solid waste management.

Integrated constructed wetland systems, in particular, are up against lack of long-term historical experience. These systems are up against a general and relative lack of technical and practical applied experience within the field. This is especially true for vertical subsurface-flow constructed wetlands. For these systems we do not have 30 to 40 years of practical, applied O&M experience to rely upon, as is the case for conventional technical options. On a more practical level, such systems are up against some specific O&M issues relating to uncertain loading rates, lack of experienced staff, technical unknowns regarding the solar-powered pumps, the vertical-flow siphon-powered distribution system, and so forth.

Technically, the designed system is risky because it entails elements not too many local people have experience with, such as siphons and solar-powered pumps. But these are calculated risks because backup systems were provided for the technical components that might run into O&M problems. Our approach for the technically riskier components was that they were important to include because they contributed to the general development within the wastewater management field. But also, that they were included in such a way that if these experiments did not work, the wastewater system as a whole would not break down.

Two specific potential weaknesses are the less-than-full coverage of both the separated collection system and the oil and grease traps. These two important issues for functionality were only partly dealt with during the construction of the system, and this might result in future O&M problems if not dealt with systematically and effectively.

Why were they only partially addressed? The hotel owner did not want to allow the conversion of the semi-covered collection system in her area, the restaurant owners did not want to have to install oil and grease traps in their kitchens, and there were no municipal by-laws to force this through. All these valid and factual reasons proved very difficult to deal with.

To these could be added the issues of construction fatigue and the normal tendency to take the seemingly easy or trivial route. Designing and constructing such a wastewater management system—in a highly complex political setting with competing national political factions that wanted to be in charge of the tsunami reconstruction funds, and competing national, provincial and municipal actors, and a donor-financed municipal infrastructure, using an innovative approach requiring lengthy discussions and negotiations, on an island located 40 km from the mainland which required all materials, gravel, equipment, and staff to be ferried to the island and remain on the island for more than one year—was an extremely time- and energy-consuming task. Toward the end of such an endeavor, construction fatigue typically sets in. Managers and workers want to finish the work and move on. This can lead to lack of energy to solve some of the trickier issues, especially the ones not purely technical but nevertheless linked to people, discussions, and long negotiations—like convincing the restaurant owners that they should make the effort to have their kitchen retrofitted with oil and grease traps.

Issues that might seem less complex or technical and therefore more trivial (e.g., oil and grease traps) are in fact often the most difficult and time-consuming, and construction projects should probably begin with them. This is especially true in the context of "construction fatigue." The Koh Phi Phi project started out with the most interesting and technically challenging element for all involved, the construction of the flower and the butterfly. The construction of and adjustments to the collection system were left to the very end. To leave the boring elements until later might be human nature but, in hindsight, is not always the best solution. However, this wastewater management system was not just built and then left to the municipality. A number of postconstruction safeguards were put in place.

Post-experiences: difficulties in motivating the municipality and activating the safeguards. It is only to be expected that wastewater management facilities will experience a number of O&M problems during their first year of operation. Some of these originate from the construction itself, and some from the always necessary adjustments and operational run-in period, especially a biological and innovative treatment facility like the one established on Koh Phi Phi. Because it was anticipated that close follow-up and adjustments would be required, five key safeguards for the first year of operation of the system were established.

1. A performance bond of 10% of the construction cost was provided by the contractor, which enabled the municipality and the donor to require the contractor to rectify, within the first year, any construction mistakes and/or omissions discovered.
2. A 3-year postsupervision contract was signed with a local expert to closely follow and supervise the technical O&M issues that would arise, and report these to the municipality and the donor for action to be taken immediately to ensure proper and efficient operation of the system.
3. As an integrated part of the total construction budget, a 5-year, 2.5 million baht ($79,000 USD) O&M budget was provided to the municipality, with bi-yearly installments of 250,000 baht ($7,900 USD). The purpose of this budget was to provide additional finan-

cial support to the municipality for system O&M; to provide the municipality with ample time for incorporating the full O&M costs into its normal municipal budget; and finally, to provide the donor with an opportunity to closely track the actual O&M of the plant and to adjust or, if necessary, to withhold budgets or withdraw from the project.
4. A contract was drawn up with local experts to provide assistance to the municipality, the local community, and the operator regarding public relations activities.
5. At the time of hand-off, a municipally controlled but community-based organization (a municipal committee) was in place to manage and operate the treatment facility. This committee, being fully responsible for all operational and financial issues for the facility, was chaired by the mayor and included municipal staff and local community representatives.

With these safeguards in place, it was expected that the municipality would be able to operate and maintain the plant without too many problems. As expected, the plant did face a number of operational issues during its first year, related to high BOD loading, high levels of oil and grease, operational problems with the solar-powered pumps, and soil in the gravel filter. These problems, if they had been managed and solved efficiently and quickly, would not have seriously impacted the system. However, and unexpectedly, this did not happen; the municipality did not actively take responsibility for management and problem solving, and the donor did not effectively activate the safeguards.

Two of the safeguards—the supplementary O&M budget and supervision by the local expert—functioned as planned. However, several post-supervision reports did list a number of issues that should have been rectified by the contractor, but no actions were taken. The performance bond safeguard was not activated within a year after commencement to make the contractor rectify the reported issues. No public relations support was provided as planned for creating awareness of the collection system and the oil and grease traps, and the municipal committee did not function as planned because the mayor did not engage actively in the issues.

The financial and technical tools are in place on this project, but time and energy are being wasted on reports and discussions that do not result in actions to rectify the observed O&M issues. This lack of action, as always, stems from a complex mix of reasons: a new group of stakeholders not being able to click; a donor obsessed with the blame game; a contractor who was not too willing to pay for the rectifications; the lack of a central key person to bring the issues to resolution; and a municipality not taking charge.

Case stories will always be stories in real time, and therefore only present a snapshot of the present situation. The real world consists of ever-evolving processes; in 1, 2, or 3 months or years from now the situation and context will have changed and so will the case story. How the Koh Phi Phi case story, like all our other case stories, will evolve from the present stage is rather impossible to predict. One can only hope the ever-changing mix of stakeholders will be able to cooperate and together find the energy, time, and resources required for the sustainable operation and maintenance of the system.

))) 6.3 Smart Technologies on Koh Phi Phi

The wastewater management system on Koh Phi Phi contains specific technologies that, to varying degrees, could contribute to better and more appropriate wastewater management systems in developing countries. We will briefly present these technologies, discuss the technical considerations, balances, and choices, and reflect upon the wider perspectives.

6.3.1 Subsurface-Flow Constructed Wetlands

The smart technology-spotting ace would see this as a rather elegant way to treat wastewater. It is beautiful; it is underground; it fits well into the landscape; it looks simple and natural. People will probably wonder: Can wastewater really be treated this way? Passing this first, visible test normally means we are headed in the right direction. The checklist score adds up. As can be seen in Fig. 6-10, the subsurface-flow constructed wetland technology scored high on most of the elements, indicating that this technology has promising potential for wastewater management systems in developing countries.

When constructing a wastewater treatment facility in developing countries, in this case on an island in the southern part of Thailand, the importance of issues such as simplicity, user friendliness, low cost, robustness, ease of operation and maintenance, low energy use, and aesthetics becomes evident. Add to this the island-specific issues of water, land, and energy scarcity,

Figure 6-10. Scoring of each the nine elements defining smart technologies for the subsurface-flow constructed wetland technology at Koh Phi Phi.

Smile, contextually appropriate; no expression, somewhat appropriate; frown, not appropriate.

and compatibility with tourism, and it becomes equally evident that the subsurface-flow constructed wetland treatment technology provides a positive solution to most of these issues.

6.3.1.1 The Technology: Subsurface-Flow Constructed Wetlands at Koh Phi Phi

The basic principle behind subsurface-flow constructed wetlands is simple: Wastewater flows below the surface through a gravel filter which, by different natural processes, treats (cleans) the wastewater. When it leaves the gravel filter, the wastewater is much cleaner than when it entered. Depending on the actual loading of the wetland, treatment rates ranging between 70% and 90% can be expected for organic matter and other pollutants such as nutrients, pathogens, and heavy metals. Treatment processes are based on emergent vegetation and basic microbiological reactions in the gravel filter. The resulting physico-chemical and biochemical processes roughly correspond to the mechanical and biological processes in conventional technical treatment systems.

Wastewater can be introduced into the gravel filter in two ways. It can enter at one end and then flow horizontally below the surface through the gravel filter to the outlet at the other end (a horizontal subsurface-flow constructed wetland). Or the wastewater can be evenly distributed over the full top surface of the gravel filter and then seep vertically down through the gravel filter to the outlet at the lower end (a vertical subsurface-flow constructed wetland). Both subsurface flow techniques were applied in The Flower on Koh Phi Phi (Fig. 6-11).

The wastewater on Koh Phi Phi is distributed to the first three lines of vertical subsurface-flow constructed wetlands through an inlet reservoir with three siphons (discussed in more detail in Section 6.3.3.1). The siphons ensure

Figure 6-11. Vertical and horizontal subsurface-flow constructed wetlands on Koh Phi Phi immediately after construction and before full vegetation.

even distribution of the wastewater to the full top surface of the gravel filter through a distribution pipe system located on top of the vertical subsurface-flow gravel filter. From the outlet at the bottom of the vertical filter, the water flows through a water level control structure into the three lines of horizontal subsurface-flow constructed wetlands.

The gravel filter is about 1 m deep and consists of three layers of different-sized gravel, and is underlain with an impermeable synthetic liner. The subsurface wetland contains both a filtering gravel medium and thin upper layers of growth medium, which support the growth of planted emergent vegetation such as heliconia and Canna lilies. The constructed wetland bottom is constructed with a slight (1% to 2%) inclination toward the outlet.

6.3.1.2 Technical Considerations, Balances, and Choices

When designing constructed wetland facilities, several technical considerations, choices, and balances have to be made. Here are three of the technical considerations and choices made on Koh Phi Phi. Choosing subsurface-flow constructed wetlands as the primary treatment technology seemed obvious due to the many advantages of this technology in the context of Koh Phi Phi: subsurface (invisible), no odor, sufficient treatment efficiency, landscaping, and simplicity. That decided, the key consideration was whether we should apply horizontal or vertical subsurface systems, or both.

Vertical versus horizontal flow? The benefit of a vertical system is that it is approximately twice as efficient as a horizontal flow system. In other words, it can treat twice as much wastewater on the same area. Because a rapidly rising demand was anticipated and land scarcity was a real issue, treatment capacity would be optimized as much as possible. But, as usual, benefits come with a cost—increased complexity in design and operation, and thus increased risk of failure.

Horizontal flow is a more reliable technology; it is better tested and very easy to operate, whereas the vertical flow system is a newer technology, less tested, and slightly more difficult to operate. The main difference is the *more complex distribution system in the vertical system*. The factor that ensures a vertical filter's higher efficiency is the even distribution of water on the whole top surface. In combination with siphons, this is achieved through multiple outlets on manifold pipes. But the sheer number of outlets presented a potential operational risk, and thereby risk of reduced efficiency. For example, outlets could be unbalanced and thus not discharge evenly. (In practice, this would be rather difficult for the operator to see, even for a top-surface gravel distribution system as at Koh Phi Phi.) Also, outlets could become blocked or the required pressure might not be sufficient to push wastewater to the farthest outlets, and so forth. These things should not happen but they can.

We decided to include both a vertical and a horizontal system on Koh Phi Phi. The vertical system's larger treatment capacity was preferred but, so as to not only rely on the vertical system, a horizontal system was included as well. Because this was one of the first constructed wetland systems in Thailand, and because its location was so prominent, the goal was to create a demonstration site for constructed wetland technology and for applying different constructed wetland technologies at one site.

To reduce the operational risk of the vertical system, a small feature was included. In the water level control structure in front of the three lines of horizontal subsurface-flow constructed wetlands, a control leveler was inserted that made it possible to raise the water level in the vertical system to 5 cm below the top of the gravel filter, thereby converting the vertical flow system to a horizontal system. Thus, a more fail-safe combined system was created.

Round versus square versus flower-shaped? Another consideration was the configuration. Six of the draft layouts made during the design phase are provided in Fig. 6-12. Each configuration had benefits and disadvantages for constructed wetlands, such as cost, effectiveness, land utilization, and landscaping.

A round shape is good for vertical systems because it provides the shortest distance from each outlet to the water distribution box, resulting in less risk of pressure loss and thus uneven distribution. A rectangular shape is good for horizontal systems because it provides better control of the flow. An uneven shape, such as a flower or butterfly shape, provides more room

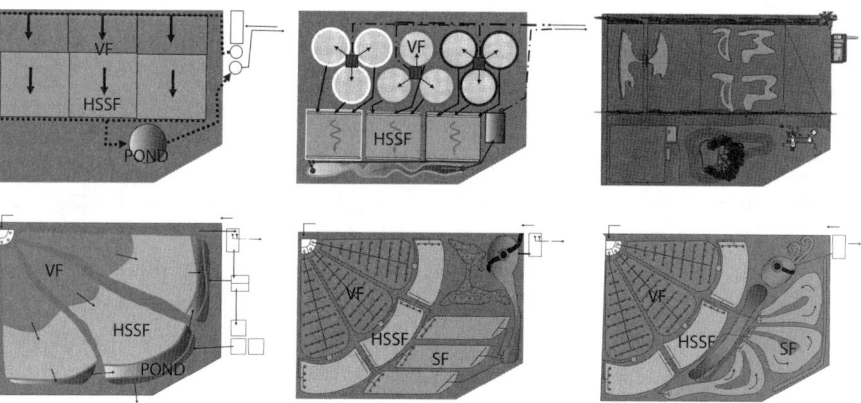

Figure 6-12. (*Clockwise from top left*) Designing the Koh Phi Phi wastewater treatment plant: conceptual, engineering, artistic, functional, landscaping, and combined options considered during the design process.

for urban integration and beautification but results in longer distances to the distribution box and less linear water flow.

A choice had to be made, and in this case the technical arguments were slightly overridden by the need for urban integration, landscaping, and beautification, but only to some extent because the flower was basically made in a rectangular form that was expected to do the major part of the treatment work (water quality was expected to be better than national standards after leaving The Flower). This allowed for more room, creativity, and odd forms in the configuration of the following treatment processes, which came to look like a butterfly.

Gravel versus soil? The media in a subsurface constructed wetland is a very important component. It was quickly agreed that the media should be available locally for reasons of affordability, replicability, and sustainability (being on an island and in a national park, "locally" still meant that more than 3,000 m^3 of gravel had to be shipped to the island—not a small task). The next issue was the type of media: gravel, soil, or a mix of these. At that time, experts in constructed wetlands were still experimenting and arguing about which type of media are the most effective.

For several reasons, we chose a pure gravel filter with three different sizes of gravel. One of the world's leading experts on constructed wetlands was on the design team and he was in the pro-gravel group, arguing strongly that pure gravel filters are less likely to clog and they have better seepage and better distribution. Also, in another part of the country a subsurface-flow wetland had failed because too much soil had been mixed into the media (saving costs), making it clog and rendering the upper layer smelly and muddy. Some members of the design team were skeptical and resistant to this choice of pure gravel, mainly because the area was supposed to look green; many questioned whether plants like heliconia and Canna lilies could be made to grow in such a gravel filter. However, the first year of operation proved that the choice was right.

6.3.1.3 General Reflections and Wider Perspectives

Is subsurface-flow constructed wetlands a proven technology? If "proven" is defined as proven to work, the answer is Yes. Experiments and pilot and demonstration projects in many different locations under many different conditions all over the world have provided enough evidence as to its efficiency. But if "proven" is defined as having been applied and successfully operated for decades, on a large scale and in all possible locations, then the answer is No. The subsurface-flow constructed wetland technology is still a newcomer in applied wastewater management, especially in developing countries.

The Koh Phi Phi design phase confirmed this. A number of international and national experts were involved in the design phase, some of them world-renowned experts within this field, and it would be wrong to say they agreed on everything. Media, shapes, inlet structures, size and type of inlet pipes, and lining—almost everything was up for discussion. This normally would indicate, in Thomas Kuhn's terms, a new paradigm still searching for the optimal, as compared to an established paradigm wherein the technology is well defined.

Today, at least three things are required to bring this technology forward and into the limelight of public and private wastewater management investments:

- More experiments in laboratories and pilot projects are needed for optimizing and coming to agreement on construction details like media, inlet structures, and so forth.
- Because this technology has proven its usefulness and appropriateness, the time has come to go from one-off projects to larger-scale implementation of subsurface constructed wetlands. A Ministry or a local government, for example, should decide that wastewater management in the country or province should mainly be based on constructed wetlands. This would move the technology out of the experimental closet into the real life of financing and treatment of wastewater for the people. Then real comparisons and improvement of factors like effectiveness, sustainability, and robustness could begin.
- This technology needs to be mixed and matched with other treatment technologies. In the Koh Phi Phi case, several different treatment technologies were applied to obtain the best results. The treatment plant functions through a combination of components that all contribute to meeting the effluent standard required for recycling the water for irrigation, including:
  - Sedimentation, anaerobic decomposition (septic tanks)
  - Skimming, oil and grease reduction (oil separator at source/pump station)
  - Filtration (screen in pumping station)
  - Sedimentation (sand trap in pumping station)
  - Filtration, biological uptake, etc. (in vertical-flow wetlands)
  - Biological uptake, denitrification, etc. (in horizontal-flow wetlands)
  - Biological uptake, etc. (in surface-flow wetlands)
  - Sedimentation and UV treatment (in ponds).

By combining and optimizing the number of treatment technologies, more efficient wastewater treatment is achieved because more processes are active compared to applying only one treatment methodology. By mixing and matching different technologies, customized treatment of wastewater with different compositions or with different discharge requirements can be achieved. Also, mixing and matching appropriate systems can allow meaningful comparisons as to cost, land requirements, and landscaping. This mixing and matching of treatment technologies is a huge area of future importance and potential.

6.3.2 Urban Integration of Wastewater Management

The wastewater management system at Koh Phi Phi is an example of active urban integration in the physical environment of the island. The system is symbolically, topographically, and programmatically interwoven in the context of the island (Fig. 6-13).

The site of the treatment facility is completely surrounded by bungalows and resorts, and along one side of the area runs the main path connecting the eastern side with the western side of the island (Fig. 6-14). With such high visibility of the facility, eye-catching, aesthetically pleasing design ideas were required. From the very beginning, the design team was up against solid resistance to a centralized wastewater system because this had already been tried and had failed (it was nonfunctioning, ugly, and smelly).

Therefore, considerable effort was spent on optimizing the design and to reveal the multiple potentials related to the design solution. A 3-D computer model was generated to present the project to the decision makers and the islanders in an appealing and accessible way, and, step by step, the general opposition to the idea of rehabilitating the municipal wastewater treatment plant diminished. This had a feedback effect on the design team, who were compelled to do their very best to meet the initial promises and implement a gardenlike park corresponding to the principles approved by the local stakeholders—two of the most important being that it should (besides work!) look beautiful and not smell bad.

Figure 6-13. Scoring of each the nine elements defining smart technologies for the urban integration technology at Koh Phi Phi.

Smile, contextually appropriate; no expression, somewhat appropriate; frown, not appropriate.

Wastewater Management Design at Koh Phi Phi   139

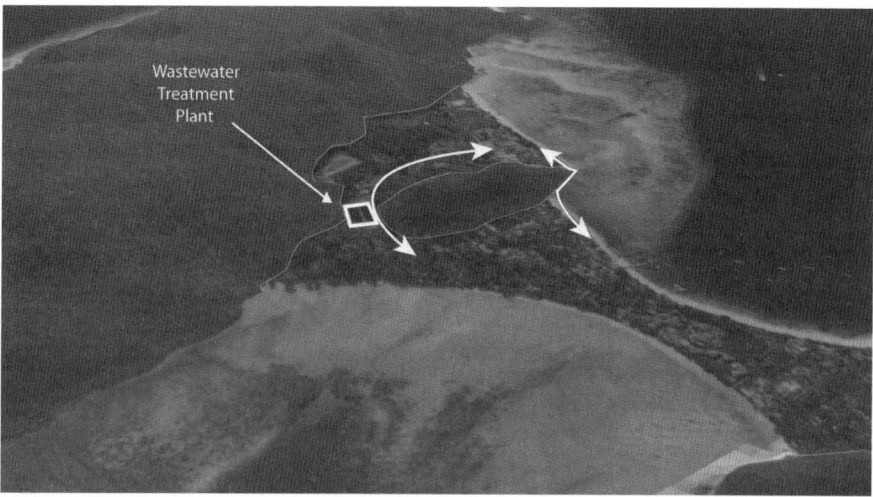

Figure 6-14. The urban layout at Koh Phi Phi.

6.3.3 Technical Considerations, Balances, and Choices

Project constraints are normal and might sometimes even be turned to opportunities. The location allowed for (actually, encouraged) an inclusive design inviting people into the area, integrating the prominent site as an asset for the island as a whole. Urban integration was essential. How was it made visible? By utilizing four different types of urban integration (see Section 6.1.2.3 for details):

- Symbolic Integration
- Aesthetics Integration (Fig. 6-15)
- Topographic Integration
- Multifunctional Integration.

6.3.4 General Reflections and Wider Perspectives

The Koh Phi Phi case combines quantitative parameters such as area demand, performance, leveling, and land availability with qualitative parameters such as aesthetics, social integration, and usability. Such combinations rely on judgments, striking a balance, and taking practical decisions. Naturally, these were all discussed and questioned, and sometimes mistakes were made. But lessons were learned and will be applied to future designs and systems. There is no doubt that wastewater management professionals will have to do much better at combining quantitative and qualitative parameters—making waste-

140 Sustainable Wastewater Management in Developing Countries

Figure 6-15. Aesthetic integration of the wastewater treatment system at Koh Phi Phi.

water facilities visible and invisible, making the visible more appealing and still effective, and more integrated while still safe.

6.3.5 Siphons

The smart technology-spotting ace would see this as a simple plastic component (Fig. 6-16). Most people would likely not know what they were looking at. The checklist score provides the same slightly mixed picture: an overall 6 out of 9 score could indicate a mixed future for siphon technology in wastewater management systems in developing countries (Fig. 6-17).

Figure 6-16. Siphon on Koh Phi Phi: a smart technology?

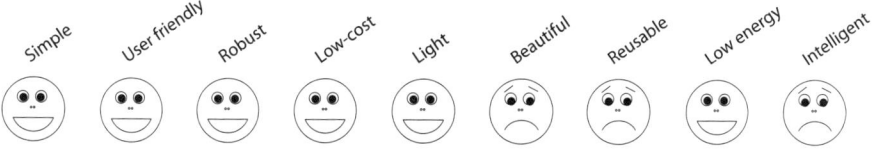

Figure 6-17. Scoring of each the nine elements defining future potential for the siphon technology at Koh Phi Phi.

Smile, contextually appropriate; no expression, somewhat appropriate; frown, not appropriate.

But the siphon scores high on issues such as simplicity, easy O&M, robustness, low cost, light weight, and low energy use—all issues of crucial importance for sustainability of wastewater management systems. It is evident that siphon technology provides a positive solution to these issues, offsetting this internal component's low scores for not being beautiful, for being somewhat difficult to re-use, or its lack of intelligence.

6.3.5.1 The Technology: Siphons at Koh Phi Phi

In a vertical subsurface-flow constructed wetland, wastewater must be distributed evenly and in intervals or pulses onto the top surface (several hundred square meters) of the wetland. A pressurized system with multiple outlet jets fed by a pump would do the job, but this would require energy consumption and pump maintenance. Because an easy, automatic, low-maintenance system with the lowest possible energy consumption was desired, especially being on an island where energy is scarce, we decided to create flush flow by using a siphon to obtain even and pulsed distribution.

A siphon is a simple mechanism that triggers the release of water when a certain water level difference is obtained between the intake and the outlet of the siphon. Once the water at the siphon intake has reached the trigger level, full flow is activated and water is discharged until the water level in the reservoir reaches a low level and the siphon starts taking in air, which stops the flush effect. Through design of the siphon, the flow rate can be determined with fairly good accuracy.

At Koh Phi Phi the siphon is connected to a reservoir with a volume balanced to the required quantity of water. The volume for each flush is approximately 8 m^3.

6.3.5.2 Technical Considerations, Balances, and Choices

Pump versus siphon. The key technical decision was whether to install pumps or use a siphon-based system. From many angles, the choice of a siphon-based system seemed obvious: no electricity consumption; no moving or mechanical parts; less odor. There were still doubts, however. Relatively few comparable siphon systems had been installed in Thailand (with success, it should be noted) and they were mostly for potable water distribution, not wastewater, and mostly on distribution systems with smaller capacities. No one on the design team had hands-on experience with the technology, and a prototype had to be imported from the United States because siphons of this size had never been produced in Thailand.

Nevertheless, the design team gave it a go and included it in the design for the following reasons: this technology had potential and should have a wider application, so the design team wanted to contribute to broader knowledge of and use of siphons. Furthermore, the cost for the siphon system was relatively low and, if it did not work, it could easily be replaced by a traditional pumping system. As it turned out, the siphon system worked effortlessly from the first day of testing and still is in operation.

The design team felt a responsibility to experiment. In a period where one paradigm has proven insufficient and others have yet to fully materialize, there is a need for willingness to experiment; otherwise, the field will not move forward. Of course, when an experiment is done full-scale, calculated risks and back-up options have to be carefully applied.

6.3.5.3 General Reflections and Wider Perspectives

A siphon-powered distribution system is an example of a simple but practical technology that reduces overall system complexity. The more often this type of simple technology can be included in an overall wastewater management system, the better its chances for sustainability.

Siphons could also be used with good effect in other types of wastewater treatment systems, for example, on-site septic tank and local irrigation systems. Why not apply it to a greater extent in on-site systems, which have some of the same characteristics and problems as the vertical-flow constructed wetland system? Wastewater leaves the septic tank in small and uneven amounts but still must reach the far ends of the underground piped drain field (this is also discussed in Chapter 4).

6.3.6 Separate Wastewater Collection Systems

The smart technology-spotting ace is a little lost here because the key feature of the collection system is that it is invisible—underground, out of sight. For the same reason, the technology potential checklist provides us with a rather mixed picture (Fig. 6-18).

Appropriateness and the future potential of collection technology can only really be understood and assessed contextually and historically. When

Figure 6-18. Scoring of each the nine elements defining smart technologies for the piped collection technology at Koh Phi Phi.
Smiling, contextually appropriate; no expression, somewhat appropriate; sour, not appropriate.

constructing a wastewater collection system in developing countries, it is extremely important that all wastewater actually reaches the treatment facility; that rainwater is kept out of the system; that leaks and infiltration are prevented; that grease and oil are kept out of the system; and that a high connection rate is achieved. A separate, closed-loop wastewater collection system provides a positive solution to most of these problems.

6.3.6.1 The Technology: Separate Wastewater Collection Systems at Koh Phi Phi

The wastewater collection system for the main business and hotel area of Koh Phi Phi consists mainly of:
- Improved connection rate and efficiency by (1) oil and grease traps for restaurants, (2) septic tanks for connected households, and (3) separate, closed-loop collection pipe systems from most households in the collection area to the pump station.
- Installation of a pump station with odor reduction features and pumps partly powered by solar energy.
- Installation of pressure pipe from the pump station to the treatment plant to avoid leakage and odor problems.

Initially, the wastewater collection area consisted of approximately 200 housing blocks and 75 shops and business, with an estimated wastewater production of 300 to 400 m^3 per day. These figures became the design criteria for the collection, treatment, and irrigation system.

The piping system consists of a main high-density polyethylene (HDPE) collection pipe with a gradient of 3 $^{0/00}$ (3 m over 1,000 m), a length of 110 m, and a diameter of 300 mm, plus five secondary HDPE collection pipe systems (200-mm diameter and same gradient) with a total length of 1,510 m.

All buildings in the collection areas had their grey and black wastewater pipes connected to the main or secondary pipes. Overflow from septic tanks and greywater was combined before being connected to the main or secondary pipe by 2-in. connections. The plan was for all buildings in the wastewater collection area to have installed septic tanks for black wastewater and all restaurants and kitchens in the wastewater collection area to have installed oil and grease traps.

The 24-m² pumping station, located on top of a damaged private wastewater treatment plant, provided mechanical treatment and a pump sump for pumping of wastewater. The pumping station contained the channel, screen, sand trap, oil and grease trap, and sump pump structure. An odor reduction system was also installed: an adjustable ventilator was installed, taking the air through a 200-mm pipe to the odor reduction box containing charcoal and wood chips.

A variable-speed control drive, an alternator, float switches for alarms, and a control panel were installed for the pumps, which meant the pumps were regulated and started depending upon actual flow to the pump station. An auto transfer switch panel was installed as an automatic changeover switch between the power from the solar panels placed on top of the pumping station and the commercial power supply from the town.

A 360-m-long, 100-mm-diameter HDPE transmission pressure pipe was laid from the pumping station to the treatment plant.

6.3.6.2 Technical Considerations, Balances, and Choices

When designing wastewater collection systems many technical considerations, choices, and balances have to be made. We addressed the following four issues while designing the collection system on Koh Phi Phi: centralized versus decentralized collection, pump versus gravity transport of wastewater, connection rates, and on-site pretreatment.

Centralized versus decentralized collection and treatment. The big question! Koh Phi Phi represents a paradox. It is a showcase presenting all the key components that should be included in a clustered wastewater management system, but it probably should *not* be copied for most locations and it definitely should *not* be copied on other islands because, normally, centralized systems are not appropriate or sustainable for small islands.

A design story: the normal mix of the chaotic, the rational, and the realistic. As mentioned earlier, Koh Phi Phi had already implemented a centralized system, financed by the national pollution control department. However, when we came to the island immediately after the tsunami event we quickly realized that this system was not and never had been working. It had a low connection rate, low or wrong gradient of the collection pipes, and a nonfunctional open pond treatment system. This meant we were back to Square 1 and had to seriously consider the possibilities of on-site collection and treatment.

But then problems cropped up with a decentralized design. The area to be serviced was the dense business and hotel district on the island, consisting of two-story business complexes, a two-story business street, a hotel complex, a six-floor hotel complex, and the main business street. This area is very compact and dense, a mix of order and chaos. Before the tsunami this area had produced almost all the wastewater on the island and had suffered relatively little in the disaster.

What else was there? The old nonfunctioning, shut-down public collection system and, to our surprise, a new private wastewater collection system servicing almost three-quarters of the buildings in the area. This system belonged to one landowner who, realizing that the municipal system would never work, had built her own private centralized collection and treatment system. The semi-covered cluster collection system was intact, well constructed, and well functioning, whereas the treatment plant had been totally damaged and needed to be rebuilt from scratch.

The remaining buildings, outside that landowner's large area, were very dense (some almost slumlike) with very limited space (Fig. 6-19). Most buildings fully covered their allocated land area and most had problems with seepage because (1) there was no land around the building; (2) there was no way to relocate septic tanks that were constructed below the buildings because no land was left for relocation; or (3) seepage-type systems did not work because they were located very close to the sea and high groundwater levels made seepage systems difficult or impossible to implement.

Figure 6-19. The dense, built-up area on Koh Phi Phi.

Now what? If a decentralized system was chosen, it could not include the primary landowner's area because she had already established and paid for a centralized cluster system. And it could not include many of the remaining buildings due to technical problems with density and seepage. With a decentralized on-site system in this specific area, only about 25 buildings could be supported.

So it was back to the drawing board. Could the well-functioning, already established, centralized but private cluster wastewater collection system be utilized? Could we create a win-win situation with the landowner wherein she would not have to pay for rebuilding the treatment systems? Would she instead provide a small piece of her land for a pumping station and allow other residents to connect to her collection system so that all the houses in the area could be serviced? Could everybody in the area be convinced to connect? Would the mayor approve of the idea, especially the notion of combining a partly private collection system with the municipal treatment plant? Things started to take shape, details were discussed and agreed upon, there were handshakes, and suddenly a design and construction project began to take form. And it was, again, a centralized cluster system.

Having conceded that it was necessary to go with a centralized cluster system, the lessons learned from the previous collection system had to be considered. Thus, the design set the following criteria for the new system:

- It had to keep the dense areas free from local seepage systems.
- It had to minimize the risks of odor problems.
- It had to prevent clogging in the collection system.
- It had to prevent sewage from reaching the streets during heavy storms.

The concept of a separate small-pipe collection system to transport grey and black wastewater from the households to a communal collection system met these criteria and was established for most connected houses. To minimize the risk of sewage reaching the streets during heavy storms, the collection system in most places was designed to separate the polluted domestic wastewater from the clean stormwater run-off.

Pumping or gravity flow? Our second important consideration was the use of pumps. O&M costs for pumping wastewater are often key factors reducing efficiency and sustainability of a wastewater management system. The number of pumps must be kept to a minimum and the use of gravity flow maximized. The treatment plant location on Koh Phi Phi was, unfortunately, located higher than the collection area, and there was no choice but to install a pumping station to bring the wastewater from the collection area to the treatment plant area.

Inclusive or exclusive of private connections and pretreatment? The third important issue was whether construction should include connecting directly to people's houses, including tertiary pipe connections. Construction of central collection systems usually includes only primary and secondary collection pipes, and most of these systems fail for exactly this reason, especially in countries with no laws or regulations forcing households to connect to common collection systems. Early on we decided that the collection system should include tertiary pipes to single houses, including excavation, pipe laying, and environmentally sound backfilling. It should also include installing septic tanks for buildings in the collection area that had none, and installing oil and grease traps in restaurants and hotels in the area. We discussed whether this was fair or correct because this would benefit only some parties, while later newcomers would have to bear these costs by themselves. In the end, we decided that tertiary connections and pretreatment were important for collection efficiency; that the cost was relatively low; and that this would impart an "accommodating" image to the system, which would hopefully reduce residents' resistance.

6.3.6.3 General Reflections and Wider Perspectives

The (partly) separated wastewater and stormwater collection systems at Koh Phi Phi are not a unique solution or the result of any new, groundbreaking research. Separate collection systems are implemented in many places but are seldom seen in most developing countries. Usually, the existing collection system started out as a drainage system. More or less accidentally it began to receive more and more domestic wastewater and slowly turned into a combined drainage and wastewater collection system, but it was never intended for or designed as a wastewater collection system.

The advisability of collecting both stormwater and wastewater in combined systems is considered every time a new system is designed. Our noncontextual opinion is that separated collection systems are by far preferable to combined collection systems. If money was no object—if it was possible to start from scratch—would clean and dirty water be mixed and then transported far away to be treated at very high expense? Mixing rainwater and wastewater creates all sorts of technical problems for managing both of them. Mixed wastewater and stormwater pollutes the streets during flooding, and mixed waters often flow in open channels, creating health risks and odor problems. Mixed waters create immense problems of dimensioning for quantity and quality at treatment facilities. If possible, rainwater and wastewater would be kept separate.

6.3.7 Solar-Powered Pumps

The picture of the solar-powered pump system used on Koh Phi Phi (Fig. 6-20) and its score on the invisible checklist indicate a paradox about solar energy.

Figure 6-20. The solar panel system at Koh Phi Phi: a smart technology?

The smart technology-spotting ace will recognize it as an interesting, smart technology, but its low score (1 out of 9) makes this somewhat more complex. Why this mixed picture?

Energy consumption is often the Achilles heel in the O&M of wastewater management systems in developing countries. Sustainable wastewater management systems must have a strong focus on using as little electricity as possible, and this is where solar-powered pumps come into the picture.

6.3.7.1 The Technology: Solar-Powered Pumps at Koh Phi Phi

A solar power system was an integral part of the pumping station in the central town area. The system consisted of solar panels, a charger controller, a battery, a backup generator, and a bidirectional inverter/charger. The system was designed to power one pump for 6 hours per day. For the other pump, and for the remaining part of the day, power would be supplied from the electrical grid on the island.

The solar panels were made of single-crystal silicon, which has a life span of up to 20 years and about 14% conversion efficiency. The total cumulative output was about 9 kW. The approximate dimension of each solar panel is 5 × 3.5 m, and each panel weighed about 70 kg. The panels were installed on top of the pumping station to avoid damage and for better positioning relative to the sun. Batteries were used for power storage to support operation during the night.

6.3.7.2 Technical Considerations, Balances, and Choices

On the surface, it seemed to make sense to reduce energy costs for the wastewater management system as well as to promote locally produced, sustainable energy. But did it really make sense? The key considerations were the complexity of the solar system and the issue of true sustainability, both of which had financial implications (Fig. 6-21).]

It is good, but also complex! The key technical decision was whether to power the pumps by the electrical grid or solar power. The choice of solar-powered pumps did in many ways seem preferable: no electricity costs, no pollution, no noise. Still, the main technical doubt came from the simplicity test. This doubt did not lessen when an electrical specialist came up with the schematic diagram of the solar power station (Fig. 6-22).

This was not exactly an iPod with all of its components hidden beneath a smooth, cool surface! Our impression was of a system with too many external components, too much that could break down—it was too complex. However, despite its lack of technical smartness, we included solar power in the system for other reasons, such as the need to experiment and contribute to developments within the field of wastewater transport. Technically, we could have considered other types of low-energy pump systems, such as the Archimedes pump.

Cheap, but only if it is provided for free! Another major concern was the investment cost and financial sustainability of the solar system. When in operation, solar panels provide an almost free source of power. The problems arise when initial investment and replacement costs are considered.

With an island kWh price of about 15 baht ($0.50 cents USD) and a daily power consumption of around 250 kWh, the annual savings would give a payback time of about 30 years (if it were on the mainland, the payback period would have been almost 100 years). Not exactly impressive. In this case, however, a donor grant covered construction investment costs, making this investment cost-free for the municipality.

The second and perhaps more worrisome aspect was the replacement costs. The life span of the batteries was about 10 years, meaning that the yearly

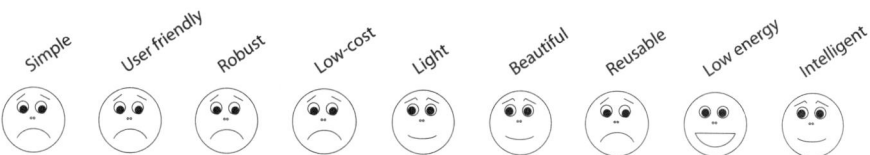

Figure 6-21. Scoring of each the nine elements defining smart technologies for the solar panel system at Koh Phi Phi.

Smile, contextually appropriate; no expression, somewhat appropriate; frown, not appropriate.

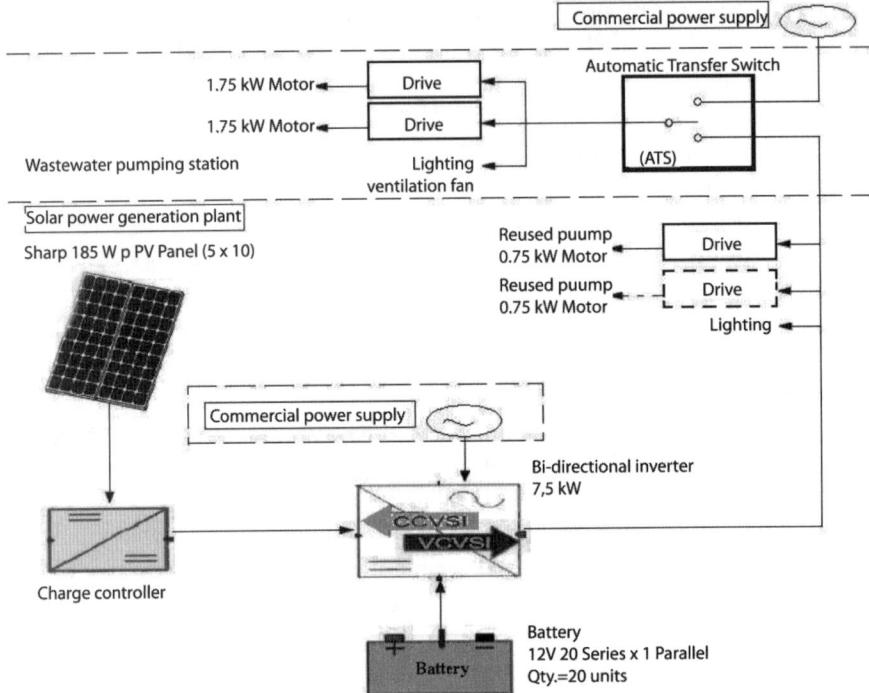

Figure 6-22. Schematic diagram of the solar panel system.

electricity savings would not be sufficient to pay for the replacement batteries. The life span of the solar panels was 25 to 30 years—less of a concern compared to the life span of the batteries. Even today, solar power at its present technical level is a nonviable financial technology for powering wastewater pumps compared to traditional methods for generating electrical power.

Nevertheless, we included solar power in the system for the following reasons: (1) solar technology has the potential to lower operation costs for electricity, and (2) the design team therefore decided it was important to contribute to the knowledge base and attract attention to the development and use of solar-powered pumps for wastewater management.

6.3.7.3 General Reflections and Wider Perspectives

The continuing evolution of solar power systems should lower their costs, increase their efficiency, create a more integrated solar and pump system, and extend the life spans of panels and batteries. Recently invented photovoltaic cells have reduced the cost of solar power to less than 10% of conventional solar panels. Although the expected operational life span for these cells is considerably less than for solar panels, the overall costs have been reduced considerably. Another development has been simpler, more integrated package systems combining pumps with solar and wind power sources.

In addition, innovative thinking begets new ways of integrating solar panels into architecture and the urban environment. The current, somewhat rigid, design of the rectangular solar panel box is now being superseded by photovoltaic cells integrated in façades, window panels, roofing materials, or free-standing sculptural elements.

Even though some doubts have been raised about the present feasibility of solar-powered pumps, it is widely believed that they will and should play an integral role in future wastewater management systems in developing countries.

7

Energy-Optimized Wastewater Treatment at Siriraj Hospital: A Large-Scale, On-Site Treatment System

⟩⟩⟩ 7.1 Year after Year

On the banks of the Chao Phraya River, just across from Bangkok's historic Grand Palace, a cluster of high-rise buildings breaks the skyline. This is the Siriraj Hospital (Fig. 7-1). Given its status as one of the best public hospitals in Thailand (and the preferred hospital of His Majesty the King), the corridors and atria are packed with patients, relatives, nurses, and doctors. The hospital has a capacity of 3,000 in-patients but the total daytime population can be up to 10 times that figure, all crammed into buildings on a few hectares of land.

Historically, the location of the hospital between the Chao Phraya River and the canal city of Thonburi provided optimal conditions for public access, water supply, and wastewater discharge. Conveniently, the hospital was built right next to the country's largest river, which could flush all possible contaminants away from the local environment. The hospital did this for many years, as did all other similar facilities, industries, and cities upstream. However, as the quality of the water in the canals and river deteriorated and Bangkok changed from water-based to land-based transportation systems, the hospital location became less convenient in terms of public infrastructures. Urban development was now concentrated along the new main roads and the hospital became more and more of an isolated island "behind" the city, making utility service to the hospital difficult and expensive. Water supply from the waterworks was costly and wastewater collection and treatment systems were nonexistent.

At Siriraj Hospital, the sterile environment of white clothing, single-use bandages, and sterilized needles that characterizes the front areas of all hospitals coexists with an equally important, but less visible, backside of indus-

Figure 7-1. Siriraj Hospital on the banks of the Chao Phraya River.
Courtesy of Ayuth Wongsomthakul.

trial laundries, barrels of hazardous waste, tons of nonrecyclable plastics, and large amounts of wastewater. Because the hospital is a huge complex with a very high population density, it is also a big polluter. In the early 1980s concerns were raised about Siriraj Hospital being responsible for long stretches of dark-colored, malodorous tentacles of pollution in the river. What if the wastewater contained carcinogens or HIV/AIDS or infectious disease pathogens? What health risks did this pollution create for the residents of Bangkok? Commuters using the river boats and neighbors of the hospital filed complaints, demanding that a wastewater treatment system be installed at Siriraj. The director of the hospital was fully aware that the consequences of inaction would be a rising public outcry, which would diminish the hospital's reputation as being one of the best in the country and associate it with unconcern for and irresponsibility toward the environment. Flushing tons of wastewater into the river, thus potentially increasing the number of customers in the hospital wards, was not a good idea. The board of directors knew that waiting for the municipal wastewater system to reach the hospital was fraught with uncertainties. So Siriraj Hospital decided to handle the problem itself.

The implemented design was an on-site wastewater management system wherein the locally produced wastewater was collected, treated, partially re-used within the hospital grounds, and/or discharged to the river. It was a self-reliant system independent of public sewers and large-scale public investments. The system was constructed and put into operation in the mid-1980s (Fig. 7-2).

7.1.1 Prescription for a Healthy Collection System

The collection system is divided into a storm drainage system that quickly directs stormwater run-off to the river, and a submerged sewer system that collects and transports the grey and black wastewater from the hospital. The sanitary sewer system is a mix of septic tanks, gravity-flow and pressure pipes, and pumping stations. For every hospital building or cluster of buildings, the wastewater is collected and pretreated in local common septic tanks (more than 100 of them are scattered around the hospital grounds). The septic tanks work as bioremediators close to the source, which separate out large, solid materials and skim off settleable and floating sludge. This improves transport through the collection system and prevents clogging. From the septic tanks, the effluent moves through a network of gravity-flow pipes to a handful of pumping stations, from where it is pumped to the wastewater treatment plant.

7.1.2 The "Paddle-Wheel Steamer" of Wastewater Treatment, Still Going Strong

The wastewater treatment plant at Siriraj Hospital is an advanced system wherein activated sludge is returned to optimize the biological treatment process. The process tanks are stacked vertically on top of each other to minimize the land requirement. In addition, the treatment system is housed in an anonymous concrete building on the hospital grounds because one of the design goals was to make the treatment system invisible.

The highlight of this system is the use of aero wheels to aerate the water and catalyze the biological treatment process. Just like an old paddle-wheel steamer,

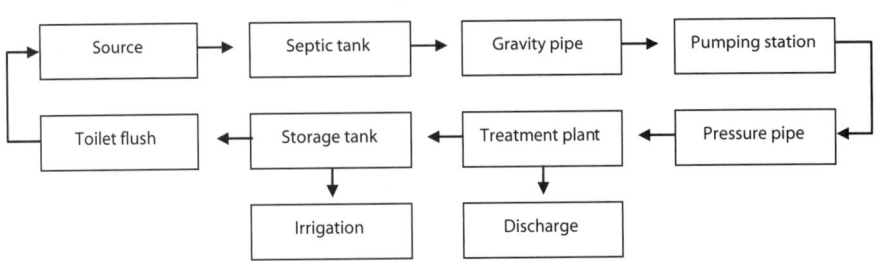

Figure 7-2. Wastewater management system at Siriraj Hospital.

the aero wheels create a steady rhythm as they rotate in and out of the water. They are coated with a biofilm of aerobic bacteria that digest organic matter when it is partly exposed to the air and partly submerged and in contact with the wastewater. The slow rotation speed (one loop per minute) creates minimal friction in the water, which minimizes energy consumption while still ensuring sufficient turbulence and the proper mix and aeration of the wastewater.

The treatment system has a standard setting that meets the mean level of pollution and average flow of wastewater per day. The slowly rotating aero wheels and their steady operation makes it a system similar to a ship that goes on and on with minimal adjustments once the course has been set (in contrast to a car, where one constantly speeds up and brakes) (Fig. 7-3). This "paddle-wheel steamer" of wastewater treatment is a simple, robust, energy-optimized technology that has stayed on course since the facility was opened two decades ago.

7.1.3 F-LUSH: Re-Using Water for Irrigation

During the first years of this system's operation, all the treated wastewater was discharged to the river. Later, financially motivated to reduce water bills, the on-site wastewater management system was extended to include local re-use of some of the treated wastewater. Consequently, some 100 m^3 a day are now redirected from the outlet pipe to irrigate a riverside park at the hospital. The grass is kept green by submerged irrigation pipes and the majority of the water re-enters the water cycle through evapotranspiration processes in the lawn. Another 100 m^3 a day is divided between two blocks of student dormitories for a toilet-flushing system. The treated wastewater is temporarily retained in storage tanks and is fed through a sand filter before it is stored in the cisterns of the toilets and urinals. There are no smells or aesthetic problems because the treated water is clear and odor-free. The facility has signs warning people not to drink the water (Fig. 7-4). To give users a choice, only half of the restroom

Figure 7-3. Aero wheels at work at Siriraj Hospital.

Figure 7-4. Re-use of wastewater at Siriraj Hospital with signs explaining "recycled water."

fixtures are connected to the pipe with recycled water, whereas the rest are connected to the conventional water supply system. No reluctance to use the recycle system has been noted.

Today, the hospital still receives about 50 visiting teams of wastewater professionals and students every year; they come to study the operation and design of the treatment plant. Twenty years after completion the plant is still an attraction and the system evokes admiration and inspiration. This is largely the result of the integration of on-site management, a robust technology, and the willingness to experiment.

))) 7.2 Reflections on Appropriateness and Sustainability

This large, full-scale wastewater management project in the complex context of downtown Bangkok was partly the brainchild of Dr. Ksemsan Suwarnarat, whose experimental wastewater systems in his private home were described in Chapter 4. Now, two decades later, what makes this system an admirable case study and an example of best practices in a tropical metropolis (Fig. 7-5)?

The light in the darkness. To find one's way in the darkness, one must look for a light; however, regarding wastewater management in the dense inner-city areas of Bangkok, few lights are to be found. Despite massive investment of more than $2.5 billion USD in centralized wastewater management systems, most of the city's wastewater remains untreated. The system at Siriraj is one of the few guiding torches in the mist of Bangkok. It intends to eventually integrate all six elements of wastewater management in a self-sustaining system, and although the appropriateness of some of the solutions can be questioned,

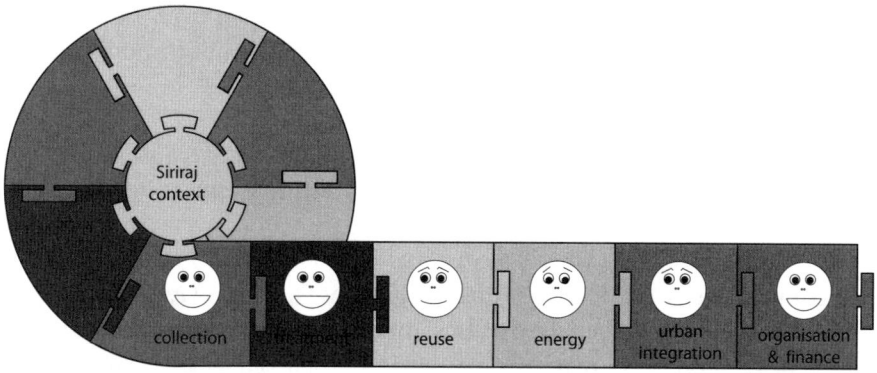

Figure 7-5. Contextual appropriateness scoring of the six elements of the wastewater management system at Siriraj Hospital.

Smile, contextually appropriate; no expression, somewhat appropriate; frown, not appropriate.

the ambition to develop promising alternatives and to suggest a better way forward is necessary and worth cherishing.

The DIY of wastewater management. Local production, local management, local financing: Siriraj Hospital represents the "do-it-yourself" mentality of wastewater management. Wastewater produced within the hospital grounds is collected, treated, and re-used, and the whole system is operated and financed by the hospital. The hospital is no wastewater burden for the municipality—it requires no public services and does not export the problem to other parts of town.

Collection. Because Siriraj Hospital is located on a riverbank slightly above the water table, infiltration of groundwater into the collection system was a significant risk that prevented the designers from burying any components very deeply. This situation initially limited the use of gravity-flow pipes, so gravity flow was used in subcatchment areas only, followed by a pumping station feeding a pressure pipe. Because the wastewater treatment facility was designed as a vertically stacked system, the water had to be pumped several meters up no matter what, which precluded a full gravity-based system. A few smaller pumping stations, located in the middle of the land plot and all feeding pressure pipes to the treatment plant, have the (somewhat cynical) advantage that they cannot be easily and unnoticeably switched off to bypass the treatment facility.

Treatment. Wastewater treatment plants based on the principle of activated sludge usually do not work in developing countries, but the system at Siriraj Hospital does. The project meets or exceeds international standards, with effluent levels of BOD below 5 mg/L, suspended solids below 10 mg/L, and total Kjeldal nitrogen (TKN) around 15 mg/L. The original design called for a trickling filter for the biological treatment, and this was approved and budgeted. However, Dr. Ksemsan, who had just returned from his doctoral studies in Germany, reviewed the initial design and was not impressed. He had a new idea. Inspired by a project he had come across in Germany, he suggested the use of aero wheels instead of trickling filters. They would be cheaper to construct and significantly cheaper to operate because of a much lower level of energy consumption. Also, it was a simple and, as it turned out, robust technology.

Although the aero wheels were based on a German concept, almost all the components were manufactured locally in Thailand. (Actually, almost all the parts that had been originally shipped from Europe broke down and had to be replaced by new, locally manufactured spares.) The drawings might have been German but the physical manifestation of the final design was completely Thai.

Organization and finance. Financially, the project is very sustainable. The O&M costs for the wastewater management system are included in the hospital's operational budget, in line with electricity, medical supplies, salaries, and so forth. The system serves only the hospital so there is no reason to make the financial system more complex than necessary (i.e., no need for tax collection, external sale of treated wastewater, or the like). As long as the board of directors maintains its stand that wastewater treatment is important for the physical environment and the public reputation of the hospital, the plant will be kept in operation.

Sustainability of this advanced systems is further enhanced by (1) in-house capacity in terms of engineers, chemists, and researchers to operate and monitor the system; (2) a plant operation training program and follow-up supervision scheme is in place; and (3) ownership and financial responsibility were determined from the very beginning. That this project was developed by a grassroots group of academics devoted to the use of appropriate technologies, in contrast to the tight-knit political, economic, and private interests that shape many other wastewater management projects, only increases its chances for continued success. Also, the fact that the Siriraj demonstration project is included the country's leading university of public health (Mahidol University), and it has a group of committed consultants and the prestigious accreditation as the hospital of the royal family, makes everyone do their outmost to ensure its success.

Re-Use. Only about 5% of the daily wastewater produced at the hospital is recycled for irrigation and toilet flushing. The remaining 95% is discharged straight into the Chao Phraya River. This is a rather modest level of re-use. The re-use system has been in operation for six years and was partly justified by the reduced water bill. In the meantime, however, the price of tap water has decreased, which has made the economic benefit of the re-use system less evident. The cost of operating the re-use system is about 40% of the cost of piped water. Taking labor for maintaining the system into account, this does not make re-used water very competitive with tap water.

Because sand slowly accumulates in the pipes, the re-use system has experienced problems with water backing up. Consequently, the flow must be manually reversed (backflushed) once a week—an annoyance to the staff. Even though it is a fairly simple task, the whole re-use issue is starting to seem illogical and unnecessary in relation to the low cost of tap water. According to the simplicity test, the re-use system at Siriraj is an additional hurdle that is not strictly necessary because the river is nearby and 95% of the treated water is flushed that way anyway. It is hard for the operators to justify why they should continue doing the laborious job for such a small amount of recycled water. The re-use project balances on the edge of being solely idealistic, which rarely supports a sustainable solution.

Energy efficiency. The energy costs of Siriraj's treatment system are only about half of that for similar advanced wastewater treatment plants. The total cost of wastewater treatment at Siriraj Hospital is 2.40 baht ($0.07 USD) per m^3, of which about 42% is spent for electricity, 4% for chemicals, and 54% for staff and other overhead costs. Still, it can be said that Siriraj Hospital has implemented an energy-efficient treatment *unit* (the aero wheel) in a energy-intensive physical *system*. Pumping the wastewater several meters to the top of the treatment plant is an energy-intensive solution. Given the physical premises, the design solution of stacking the treatment facility was justified, but from an isolated energy point of view was inappropriate.

Physical urban integration. Both the riverfront garden irrigated by treated wastewater and the invisibility of the wastewater treatment plant among clusters of hospital units are aspects of urban integration from an aesthetic point of view. In addition, the choice of a compact and area-optimized treatment system is a function of the high land costs in central Bangkok and the limits on horizontal expansion, which forces property owners to densify heightwise.

⟫⟫ 7.3 Smart Technologies at Siriraj Hospital

The aero wheel, also known as a *rotating biological contactor* (RBC), a *submerged contact biodisc aerator* (SCBA), *rotating perforated tubes* (RPT), or a *pipe biofilm reactor*, is an example of an applied technology that includes several elements of appropriateness and sustainability, and thus calls for a more detailed presentation.

7.3.1. The Aero Wheel

The aero wheels at Siriraj Hospital were based on the concept of the Stella Magic from Germany. It follows the design of RBCs that were first installed

Figure 7-6. Aero wheels for large-scale, on-site treatment: a smart technology?
Smile, supportive element for overall potential; no expression, somewhat supportive; frown, not supportive.

in Germany in the 1960s and which have since been developed and refined into a range of reliable and robust operating units now on the market. The aero wheel has been successfully implemented in numerous cases at hospitals, hotels, and shopping centers and is continuously gaining footholds in various parts of the world (Fig. 7-6).

The aero wheel is a technical subcomponent in an advanced wastewater treatment system that provides oxygen to the aeration tank and allows for aerobic biological digestion (Fig. 7-7). The aeration takes place as simple hollow cylinders mounted along the perimeter of the aero wheel rotate in and out of the water. The aero wheel utilizes at least four simultaneous processes that contribute to the aerobic treatment, including: (1) air diffusion/dissolved oxygen/suspended bacterial growth; (2) attached growth/fixed biofilm; (3) trickling/flushing; and (4) mixing.

As the wheel rotates, atmospheric air is trapped in the tubes and is transported to the lower parts of the aeration tank, where the design of the tubes

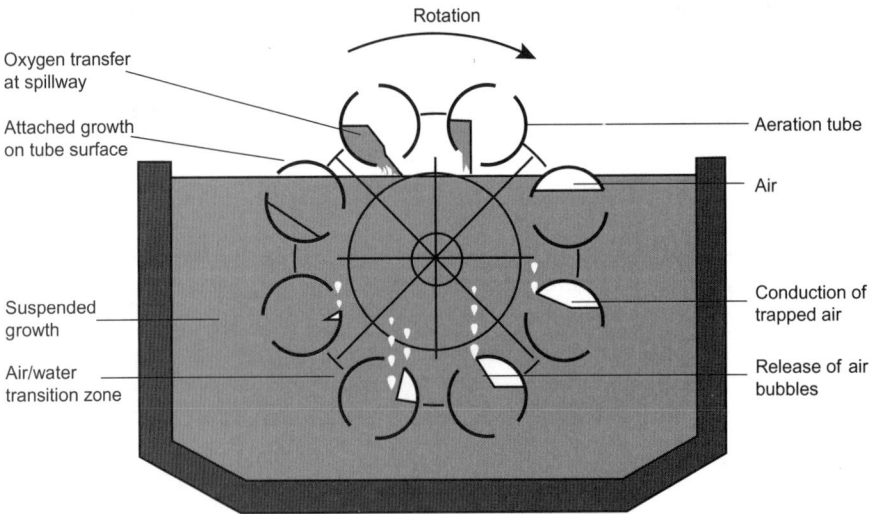

Figure 7-7. Aeration by aero wheels.

allows the trapped air to diffuse into the water. In the turbulence caused by the rotation of the wheel, oxygen is dissolved in the wastewater, creating optimal conditions for suspended aerobic bacterial digestion. As the air leaves the submerged tubes, they are consequently filled with wastewater which is transported to the surface of the tank. There it absorbs additional oxygen when it is flushed or trickled out of the tubes, thus providing space for more air and a new cycle of aeration. In addition, microorganisms settle on the tubes and grow with the help of air trapped under the water and in the atmosphere, alternately breathing and feeding with each revolution.

The aero wheel, which has a diameter of about 3 m, typically revolves at one rotation per minute. The slow rotation speed reduces friction between the tubes and the water, which reduces energy demand. When comparing energy consumption per cubic meter of treated wastewater, this method only requires about 5% of that needed by conventional aeration devices such as aerators, impellers, or diffusers.

This system can be adjusted for constant operation and performance to meet the required mean levels of pollution, effluent criteria, and the average flow of water. Through steady operation, minimum O&M activities are required.

7.3.1.2 Technical Considerations, Balances, and Choices

The fact that the aero wheel can aerate wastewater at one-twentieth the cost of conventional aerators means an approving nod is in order. It uses more energy than do ponds and constructed wetlands but, given the very limited land available in the central parts of Bangkok, it makes good sense. At Siriraj Hospital the trade-off between land use, energy consumption, and the availability of skilled staff from the hospital and university makes the aero wheel a sustainable and appropriate technical solution at that specific location.

The system is an example of successful transfer of technology because today a local market has developed for the manufacture of aero wheels and spare parts, thereby ensuring this system's continuous operation and maintenance. If this had not been the case, the implementation of an imported blueprint technology such as the aero wheel could easily have become a technology difficult to sustain.

7.3.1.3 General Reflections and Wider Considerations

Given its treatment efficiency, low land area demand, low energy consumption, and low unit cost, the aero wheel and similar RBC systems have potential as a reliable and energy-efficient wastewater treatment system suitable for serving high-rise buildings, commercial complexes, hotels, hospitals, and universities in areas with limited land availability. The aero wheel also broadens the potential for developing and installing on-site treatment plants, designed and fabricated by modern industrial methods, where high-tech facilities can be operated and maintained by skilled staff. These wastewater treatment units can be robust and can be implemented as a modular system, enabling gradual extensions along with increasing quantities of wastewater.

A concern, however, is the amount of sludge produced. A consequence of aerobic biologic digestion is a relatively large production of sludge that must be removed from the tanks and managed in an appropriate way. In dense urban settings, local re-use and management of sludge is rarely an option, so an evaluation of the overall appropriateness of the system always requires an assessment of how the collection, transportation, re-use, and incineration or burial of the sludge will be managed. In the case of Siriraj Hospital, surplus sludge is managed in collaboration with the municipal wastewater management system (i.e., it is deposited in a specially designed cave on the outskirts of town).

8

Constructed Wetland at Patong: A River Treatment System

))) 8.1 Doing the Next Best

In Thailand, Phuket Island and its city of Patong are a major international holiday destination. A sparkling nightlife and great seafood may entice visitors already there, but what draws tourists to Patong municipality in the first place is its stunning white curving beach facing the Andaman Sea (Fig. 8-1). It has undergone extensive reconstruction after the devastating 2004 tsunami and tourists are flocking to the area once again.

But there is another potential danger from the water that could have a negative effect on tourist visits—wastewater. Only some of the municipality's wastewater is treated in wastewater treatment plants; the remainder is discharged directly to streams and rivers leading to the sea.

Three factors make the situation in Patong particularly serious. First, Patong is located in an enclosed area and all of its canals merge together with only one outlet into the sea. All wastewater produced in Patong therefore ends up flowing out the Pak Bang River into Patong Bay. Second, the crescent shape of Patong Bay makes the discharge stay longer in the bay than in, for example, the straight Bay of Karon farther south on the island. And finally, Patong is highly dependent on tourism and therefore on the quality of its beach water.

Taking care of tourism is of paramount importance. It creates jobs and income, and generates taxes. But in recent years the international tourism industry, as well as local governments, have become more aware of the fact that tourism and taking care of the environment go hand-in-hand. As stated in 2005 by Mr. Pattanapong Aikwanich, the president of the Phuket Tourist Association, "We cannot separate one from the other. If we do not clean our beaches in Phuket, the tourists will boycott us. If we do not control housing, provide clean and safe environments for our guests, get rid of solid waste properly and treat wastewater from rivers accordingly, we will be out of business in no time. Our hotels have received complaints about smell from sewers,

Figure 8-1. Patong.

Source: Google Earth, with permission.

pictures taken by tourists of black water entering Patong beach and plastic bottles and other type of garbage coming in from sea at the end of the monsoon season. We need to tackle challenges like that."

8.1.1 Going to the Beach

Patong has an existing wastewater management system with sewers and an advanced wastewater treatment plant, which is in operation and well maintained. This system, however, is inadequate because only about 50% of the wastewater in Patong is collected and treated. The remaining wastewater (greywater as well as effluent from septic tanks) is discharged to the drainage system, primarily the two major waterways in the town: Pak Bang River and Pak Lak Canal (Fig. 8-2). These merge upstream and then transport all stormwater and wastewater to the estuary right at the beach of Patong, and then into Patong Bay.

The water quality of the river is poor, with BOD levels above 20 mg/L. Also, the level of pathogens in the river water is very high. The polluted river water therefore creates not only a public health risk but also enriched levels

Figure 8-2. Pak Bang River upstream and Pak Lak Canal.

of nutrients, which increase the growth of seaweed and occasionally create a thick green sullage on the surface of the water along parts of the shore—not exactly matching the image the tourists see in their glossy tourism magazines when they are about to book their holiday.

For Patong to maintain its primacy as a major tourist destination (in competition with Bali, Fiji, Hawaii, and the Caribbean), it is essential to prevent untreated wastewater from circulating in the bay. This is why in 2005 the municipality did not hesitate to launch a rehabilitation project of the wastewater management system in the town, supported by a grant from the Danish government.

8.1.2 Reducing the Problem

The basic goal of the project was to reduce the problem (i.e., improve the water quality of the river that would otherwise continue to deteriorate the beach environment). The need for improvements was urgent. One option was to call for an ideal solution—getting all the households in the town connected to a collection system that could transport the wastewater to an extended municipal wastewater treatment plant. Another ideal option was to force householders to implement efficient on-site systems. These options, however, would either be very costly, take several years to implement, or, due to lack of municipal enforcement and the number, location, and density of the households involved, would be difficult to implement.

The tourists could not wait years for a better beach environment; neither could the tourism industry or the municipality. Because action had to be taken, focus turned toward second-best solutions, without restricting or compromising the possibility of implementing the best solutions in the future when funds, adequate planning, and implementation schemes were in place. With the urgency and limited funds at hand, we (the project team) decided on a responsive solution that pragmatically identified and minimized the problem.

As a result, the agreed-upon plan was to collect polluted water from the two major canals, treat it, and discharge treated water back into the canal downstream. A municipal pond previously used for primary sedimentation of wastewater had become obsolete after completion of a new section of the wastewater treatment plant, and a pond area could be allocated for a canal water treatment system. The pond had a surface area of 6,300 m^2 and a depth of roughly 3 m.

In this relatively large area it made sense to choose a system that capitalized on the warm and sunny climate and was tailored to reduce organic matter, nutrients, and pathogens in the polluted river water. The scheme created a combined pond and constructed wetland system consisting of an inlet pond designed to remove sediments and particles, three parallel horizontal subsurface-flow constructed wetlands, followed by a settling and maturization pond prior to the outlet (Fig. 8-3). The focus was on creating a simple and robust solution.

Because this area was located close to the sea, the water level in the canals changed by up to 3 m according to changing tides, allowing intrusion

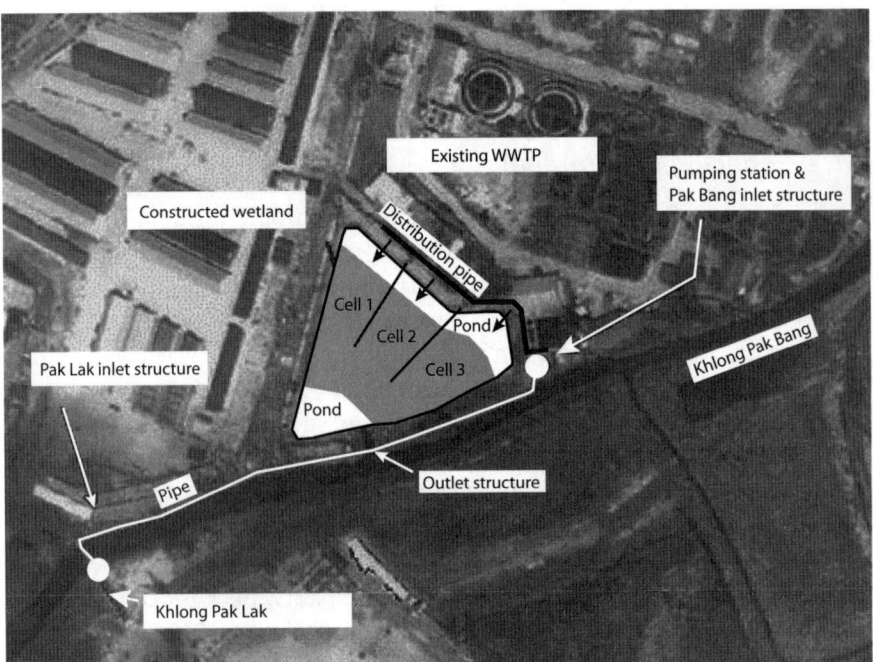

Figure 8-3. Location and layout of the horizontal subsurface-flow constructed wetland facility.

of saline water into the canals during high-tide periods. It became important to design the system in a way that rising water level would not flood the wetland system; that the plant would not treat saline seawater; and that the treated water would not return into the plant as backwash.

Because the wetland was to be constructed in an old pond, an existing embankment was already in place and protected the surrounding area from flooding. But at the same time, this necessitated a pumping station to collect the water and lift it up into the treatment system. It was not possible to have a full gravity-based system because that would allow seawater to enter the system.

The final design of the inlet structure allowed the pumps to operate only during low tides, when the water was predominantly wastewater from the town. The pumps were equipped with salinity meters that switched off the pumps if the level of saline seawater became too high. This was crucial because pumping seawater would not only damage the pumps but also reduce the treatment capacity of the gravel filter. Due to the tides, the pumps could only be in operation for about 10 hours per day. Consequently, the capacity of the pumps was dimensioned to pump large amounts of water into the treatment plant in a short period of time, filling the ponds and letting the water trickle through the filter in a sequencing schedule.

8.1.3 Design of the Inlet Structure

The inlet unit to collect water from Pak Lak Canal consisted of a number of simple components (Fig. 8-4). First, to ensure that the polluted water from the river always flowed into the inlet (including during low tides), a 10-cm-high concrete wall was built across the canal. Second, to prevent larger solids from entering the inlet, a cap unit with drain holes was installed covering the full area of the inlet. And third, water was transported from the Pak Lak Canal inlet to the pumping station at the bank of Pak Bang River via a 250-mm-diameter gravity HDPE pipe located at the bottom of Pak Bang River (Fig. 8-5).

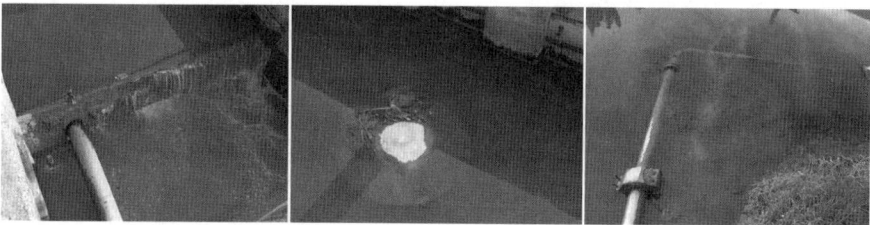

Figure 8-4. Pak Lak inlet structure (*left to right*): damming the water, inlet cap, and HDPE pipe on the river bottom.

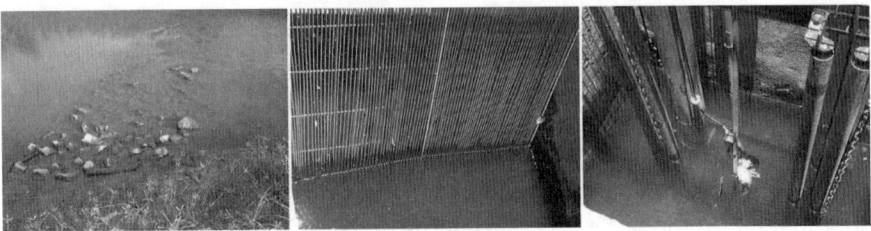

Figure 8-5. Pak Bang inlet structure (*left to right*): damming the river, screen, and pumps.

The inlet unit to collect water from Pak Bang River consisted of eight simple components:
1. A stone dike connected to the inlet to ensure that the polluted water from the river flowed into the screen and pumping station during low tide.
2. Screens to prevent garbage from entering and blocking the pumps.
3. A sand trap located behind the screen to collect larger particles, thereby preventing sand from damaging the pump impeller.
4. Four pumps to lift water from the pump sump through a pressure pipe into the inlet pond: two pumps for Pak Lak Canal, each with a capacity of 72 m^3/hour, and two for Pak Bang River, each with a capacity of 108 m^3/hour.
5. Check and gate valves to prevent the return of water from the main pressure pipe and to allow maintenance of the pumps.
6. Float switches in the pump sumps to avoid dry running if the pumps became empty.
7. Salinity meters to automatically start and stop the pumps. A salinity set-point (e.g., 3 ms/cm) ensures that the pumps stop operation if the salinity in the water gets above the set-point. Once an hour a timer lets the pumps run for 5 minutes. If the salinity meter within these 5 minutes registers a lower salinity than the set-point, the pumps continue to operate. If not, they stop. The set-point for the salinity meter controls the amount of river water the pumps can take into the wetland. The set-point is continuously adjusted to optimize the flow.
8. Hour meters to monitor the hours of operation of the four pumps.

8.1.4 Design of the Treatment and Outlet System

The three main components of the treatment system were: (1) Three inlet ponds created to remove suspended solids and sediments from the wastewater before it was led into the gravel filter. These inlet ponds ensured an even distribution of water into each of the three gravel filters. (2) Three gravel filters filled with 0.8 m of gravel (diameter 3 to 10 mm) to treat the wastewater. The water flowed evenly in a horizontal flow through the gravel filter 2 to 4 cm below the surface of the gravel. Vertical pipes were installed to allow for inspection of the water level in each gravel filter. The gravel filters were planted with Canna lilies and heliconia to enhance the treatment process and make the treatment plant more beautiful. The water level in the gravel filters can be adjusted with an overflow V-notch. (3) An outlet pond to remove sediments deriving from the gravel filter (Fig. 8-6).

The outlet was designed with a nonreturn valve so the water from the canal would not enter the system during high tides. At the same time, the outlet allowed for an overall adjustment of the water level in the wetland system.

The total capacity of the wastewater treatment plant is defined by the area available and the mean levels of contamination. Working backward from these data makes it possible to calculate how many cubic meters of water can be taken in per day and approximately how many corresponding population-equivalents (PEs) can be served. The new plant treats about 2,000 m^3 of wastewater per day, which extends the municipal wastewater management service to include another 15,000 people at a construction cost of about $25 USD per person served.

))) 8.2 Reflections on Appropriateness and Sustainability

The constructed wetland wastewater management system implemented in Patong is in many respects a special case because it treats polluted river water (Fig. 8-7). Therefore, we can assess its appropriateness and sustainability on two levels: the system as a whole, and the individual elements of the system.

Figure 8-6. (*Left to right*) Inlet pond, gravel filter before being planted, and outlet pond.

Figure 8-7. Contextual appropriateness scoring of the six elements of the wastewater management system at Patong.

Smile, contextually appropriate; no expression, somewhat appropriate; frown, not appropriate.

Is it appropriate to treat polluted river water? A constructed wetland system was implemented in Patong that provides wastewater management equivalent to 15,000 residents and it reduced (although did not eliminate) the problem of untreated wastewater reaching the tourist beach, and this was done in a cost-effective manner.

Capitalizing on the fact that wastewater was already flowing in the drains and canals of Patong reduced the complexity of the task to be undertaken, and that made sense. Often, the installation of sewers accounts for about three-quarters of the costs related to the construction of a wastewater management system, so implementing a system with no sewers created good value for the money. But still, does it make sense to treat water from a river?

Consider Taiwan, a country with broad experience in river treatment. A number of demonstration projects using different technologies have been installed there. The systems are in operation but the treatment comes at a very high cost with little impact, at least for the Taiwanese paying for the systems. Only a low percentage of all the wastewater is treated and treatment takes place outside the cities just before the rivers enter the sea. As a solution for wastewater problems on Taiwan, which includes public health risks, odor, inferior local water quality, and poor aesthetics of visibly unclean water in the inner cities, river treatment certainly does not address most of these problems. River treatment seems to have been established more as demonstration of at least doing something, not as a serious attempt to tackle the root causes and the wastewater management problems that directly affect people. For that, only effectively managed on-site or cluster systems suffice. River treatment by-passes root causes; its linear methodology lacks the circular, ecosystem approach to appropriate and sustainable wastewater management.

The same may be said for the Patong system. Phuket Island's tourism assets—its beach and Patong Bay—provide some justification for river treatment at this location, but not as a sustainable and general solution to wastewater problems in Patong. Because only a minor percentage of the total polluted river water is treated (most is discharged diluted during high tides or by-passes the pumping station during low tides), the visible impact is small and only an environmentally conscious municipality (or mayor) would maintain such a system.

The appropriateness of the individual elements. The design of the wastewater *treatment* system is robust and simple. The constructed wetland technique is robust, tested, and appropriate for the type and amount of wastewater to be treated. It is, furthermore, cheap and easy to operate and maintain, and should pose no problem for the municipality to operate together with the two large, advanced activated sludge treatment facilities on the same location. The saline-based inlet structure, however, may be too complex and time will tell whether the right choice was made.

For topographic reasons the system did not include a wastewater *collection* system, nor was a *re-use* program implemented (the latter due to the municipality's lack of incentive because a neighboring advanced facility provided daily access to 15,000 m^3 of treated wastewater, which was already being used for urban irrigation). There were also financial considerations—a fixed available budget.

Concerning *energy consumption*, the system relies on pumping river water into the wetland and then using gravity flow to push the water through the treatment system and out through the outlet structure. No solar energy, Archimedes screw pumps, or other energy-saving methods were applied. In a city with plenty of cheap and readily available power, at a location where two large, advanced treatment plants were operated by the municipality and where the electricity costs for the wetland system were almost invisible on the municipal wastewater electricity bill, it made little sense to implement costly energy-saving systems.

In terms of *urban integration*, the wetland does create a large plateau of colorful flowers and forms a beautiful foreground to the soft curves of the distant mountains. Moreover, the flower field and the landscaped banks of the wetland system preserve a large, contiguous plot of municipal land in the town. It is currently a large front yard for the four-story apartment blocks surrounding the site but, if necessary in the future, it could be utilized for other public functions. Visitors barely notice that a wastewater treatment is at work. The gravel filter is invisible, topped as it is with plants and flowers.

The original design included a public walking path and benches along the embankment separating the canal and the wetland system, and elevated paths designed as wooden bridges connecting through the area and crossing the wetland cells. However, because the area doubles as a garage and storehouse for the municipal engineering department, it was in the municipality's best interest to reduce public access to and around the site to prevent loss of property. Nevertheless, the facility achieves a satisfactory level of urban integration. The goals of full integration of landscaping, multifunctionality, and public access might not always make sense.

The constructed wetland system is located in one of the richest municipalities in Thailand—a municipality that already operates two large, advanced treatment systems on the same location. The question of sustainable *organization and finance* should therefore, in theory, be a minor matter. The municipality clearly has the capacity to operate and maintain the system. Furthermore, to be on the safe side, the donor's construction contract included a 3-year O&M support budget; technical backup from an experienced, local constructed-wetland expert; technical O&M training of municipal staff; and development of an easy-to-follow O&M manual for the system.

The municipality collaborated well with all parties during construction. Competent administrators, combined with diligent politicians who were fully aware of the interrelationship between the state of the physical environment and the economy of the tourism industry, paved the road for a smooth project. The wastewater system was inaugurated at a large ceremony with more than 300 participants, fireworks, speeches, displays of informational materials, and publication of a booklet—all arranged, paid for, and promoted by the mayor and the municipality. Finance, organization, attention, simplicity: all the key elements of sustainability were seemingly present. Nevertheless, doubts about the system's long-term viability are surfacing. The treatment system, a year after completion, was being operated and maintained but only barely. The mayor said the municipality was over its head in constructing the huge and expensive third phase of the centralized advanced system.

The future of the system? As stated by the mayor, it may continue to be operated efficiently and as planned. Such operation would be made much easier by a planned water gate to be installed near the river mouth. Or it may be abandoned, or it may be converted into high-rise buildings because its location is right in the middle of the Patong bowl where land is very valuable. Or it may be converted from river treatment to polishing the effluent from the adjacent advanced treatment plant. The first and last option would satisfy the original goal of the plant, which was to reduce wastewater problems and health risks for the citizens of Patong and the millions of tourists visiting the city every year.

))) 8.3 Smart Technologies in Patong

8.3.1 Horizontal Subsurface-Flow Constructed Wetlands

At Patong a horizontal subsurface-flow constructed wetland system treats the river water. The technology is a robust system that efficiently removes biological matter from the wastewater (organic matter and nutrients are the prime reason for algae growth on the beach). In addition, the system is low-cost (about one-fifth the cost of implementing an activated sludge wastewater treatment system) (Fig. 8-8).

8.3.1.1 The Technology: Horizontal Subsurface-Flow Constructed Wetland

Basically, a horizontal subsurface-flow constructed wetland works like a bathtub filled with gravel (Fig. 8-9). Opening the tap, water is let into the bathtub until the water level is a few centimeters below the surface of the gravel. The drain in the bottom of the bathtub, located at the opposite end of the inlet tap, is equipped with a vertical, movable L-shaped pipe that can make the water level in the bathtub exactly the same as the top of the L. By adjusting the height of the outlet pipe, the water level in the gravel filter can be controlled. The water is changed by opening the tap and making new wastewater push the retained water though the drain and out of the outlet. This creates a sequential treatment, a "biological lung" continuously inhaling and exhaling water.

In the Patong plant, the inlet system was dictated by the changing tides of the river. On high tides, seawater enters the river. As the tide changes, water is flushed back to the sea. On low tides, wastewater from the city dominates the river water and creates a steady flow of sewage to the sea. This is the time when the system is designed to take in all the wastewater the wetland can absorb, about 2,000 m³ in 10 hours. Heavy-duty pumps do this job and transfer the wastewater to a pond.

In Patong, the horizontal subsurface-flow constructed wetland consists of three parallel wetland cells. Each cell measures about 30 m × 70 m and

Figure 8-8. Constructed wetland for river treatment: a smart technology?

Smile, supportive element for overall potential; no expression, somewhat supportive; frown, not supportive.

Figure 8-9. Section of horizontal subsurface-flow constructed wetland.

has a depth of 1 m. The cells are sealed from the ground by a 3-mm HDPE membrane, which has been heat-molded at every section and is entirely waterproof.

At the inlet, outlet, and on the bottom of each cell is a 15-cm-thick layer of rough gravel (5 to 10 mm diameter). The rest of each cell is filled with fine gravel (3 to 4 mm diameter) to a depth of 60 cm. Each cell is topped with a 10-cm-thick layer of medium-rough gravel (~5 mm diameter). The larger gravel diameters at the inlet, top, and bottom prevent the system from clogging because excessive biological growth will take place at these locations. Moreover, the larger gravel size ensures more space between the stones and facilitates a more even distribution of water throughout the width of each cell.

During construction, great care was taken to not let any silt or sand into the gravel filter, to not compact the layers, and to not mix the different types of gravel because this would have contributed to the risk of clogging and reduced the gravel filter's treatment efficiency.

The biological treatment takes place as microorganisms settle on the gravel and feast on the biological matter in the wastewater (it is primarily organic matter that causes anaerobic conditions, malodorous environments, visibly black water, and incontrollable growth of seaweed and algae in the sea). The process reduces the number of contaminants by up to 90%.

To improve treatment efficiency and beautify the area, the gravel filter in the horizontal-flow constructed wetland in Patong is planted with Canna lilies. The plant is a local biotope, resistant to high concentrations of pollutants in the water and, through millennia of evolution, has been optimized for a semi-aquatic environment in the tropics. The plants contribute to the treatment process by the uptake of nutrients and water through their root systems. Carbon, phosphorous, and nitrogen are absorbed into the plant cells as the plant grows bigger and bigger. Wastewater is taken up and, through plant transpiration, is vented into the air.

8.3.1.2 Technical Considerations, Balances, and Choices

A system with or without plants? The constructed wetland experts involved in the design strongly favored the use of plants in the wetland system, and it was decided to cover the gravel filters with Canna and heliconia plants. Because the system is located within the municipal wastewater treatment plant premises and is restricted from public access, the question of beautification became less important. In the case of Patong, the wetland is a monofunctional system designed to only treat river water. No effort was made to integrate multiple synergistic potentials such as public recreational parks, commercial gardening, or a visual landmark, as was seen in the projects at Baan Pru Teau (Chapter 5) and Koh Phi Phi Island (Chapter 6). However, because the constructed wetland was the first of its kind on Phuket Island and many visiting groups were anticipated, colorful and everblooming Canna lilies were chosen as the primary plants (Fig. 8-10). Cannas grow wild all over the country, along canals, rivers, and lakes (including on the banks of heavily polluted drains and open sewers), providing clear evidence that this sturdy plant can tolerate even high levels of pollution.

During construction many people, including contractors and municipal staff, found it difficult to believe that these plants could survive in the gravel filter, especially because the constructed wetland is topped by relatively large-diameter gravel. Close supervision during construction was necessary to keep people from adding a top layer of soil to "help" the plants or placing too much soil around the plants when planting them. Too much loose soil would have clogged the system.

In Patong the plants were planted exactly as they were supplied from the nursery—rooted in small organic peat pots—thereby keeping the amount of soil to an absolute minimum with little impact on the functionality of the wetland system. The 30-cm-tall plants, with two shoots per pot, were placed in grids of about 50 cm × 50 cm (four plants per square meter). The roots of the plants expanded very quickly as they reached the nutrition-rich water

Figure 8-10. Constructed wetland (*left*) and inlet pond (*right*).

right below the surface of the gravel filter, and within a few months the gravel filter was fully covered with 1- to 2-m-tall blooming Canna lilies and heliconias (the latter not quite as fast-growing).

Determining the size of gravel. Patong is located in a bowl facing the sea and surrounded by hills. Because there were no quarries in the immediate vicinity, the constructed wetland media had to be transported from a distance, making it fairly expensive. Because no cheaper media alternatives could be found and there were several quarries on the island itself, the decision was made to use gravel. Textbook design guidelines helped determine the best size of the gravel media but, in this case, the team's constructed wetland expert went around to the stone quarries and selected the actual gravel sizes based on the current production methods and stocks available. When designing an appropriate and sustainable system, the choice of media (and ultimately the choice of the wastewater treatment system itself) must take advantage of the natural resources available in the local area, and costs must be kept as low cost as possible (e.g., by minimizing expensive transportation of gravel to the site and avoiding the use of gravel sizes not readily available in the area).

8.3.1.3 General Reflections and Wider Considerations

There is an urgent need around the world, but especially in tropical and developing countries, for low-cost, low-tech wastewater solutions that employ natural systems such as horizontal subsurface-flow constructed wetlands, to address environmental challenges. In doing so, there is enormous potential for the creation of green urban oases: creating beauty, wildlife habitat, healing landscapes, and generating useful products from the water and nutrients currently misnamed "wastewater."

The issues of public health risks and the release of nutrients causing eutrophication, and a wide range of environmental problems, are certainly not unique to Patong. Many coastal cities in developing countries are facing coral reef decline, oxygen depletion, fish kills, ecological degradation of rivers, lakes, and beaches, and giving competitive advantage to weed species over native plants in ecosystems impacted by release of human wastewater.

But the past several decades have also produced new solutions to these dilemmas, stemming from a fundamental change of perspective based on a total ecosystem approach that treats wastewater as a valuable source of nutrients and water upon which ecologically flourishing wetlands can exist. Wetland scientists have demonstrated that not only natural but also properly designed and constructed man-made wetland ecosystems are extremely efficient at utilizing and cleaning such nutrient-rich waters.

The new disciplines of ecological engineering and ecotechniques, upon which the Patong system is based, seek to utilize predominately natural, ecological mechanisms (Nelson and Tredwell 2002). This wetlands approach turns out to be easy to maintain and efficient in turning what was previously waste into green plants and re-usable water. Furthermore, wetlands are cheaper to construct and operate because there is less reliance on complex technology, which is capital- and maintenance-intensive, and they use much less electricity and fuel. The use of ecologically constructed wetlands for wastewater treatment relies on the transformative ability of green plants and nonpathogenic microbes, rather than expensive machinery. In addition, designed wetlands create a buffer ecosystem between human economy and the environment to mitigate negative impacts.

9

Pond and Constructed-Wetland Treatment at Sakon Nakhon: A Sustainable Municipal System

))) 9.1 Fields of Action

Covering 32 km², Nong Han Lake is home to thousands of birds and fish. Thailand's largest inland lake, it has long been a source of natural beauty and an abundant source of food and water for the people dwelling along its shores. In this pristine landscape—the broad plateau, the great open sky, and the wide fields dissolving into the horizon—one of the lakeside settlements developed into a lively and colorful city, a provincial capital that is home to about 50,000 people—Sakon Nakhon (Fig. 9-1). To this day, Nong Han Lake has been of great value to the people of Sakon Nakhon as a water supply reservoir, as a tourist attraction, and as a recreational and fishing locale.

But as the city has depended on the lake, the growth of Sakon Nakhon has also damaged the natural ecosystem. This lake is located at the lowest point of a regional watershed. Consequently, water is trapped in the lake with little chance to continue the flow to other water bodies, which makes the life cycle of the lake depend on high evaporation rates. Luckily, Nong Han Lake is very shallow and has a very large surface area compared to its volume, which facilitates high evaporation. Unfortunately, as the water evaporates, the relative concentration of nutrients, salts, and other pollutants increases, and year-by-year these contents accumulate. This inland lake is thus a sensitive ecosystem that is highly vulnerable to pollution. Remedial action was justified and was initiated in 1993 by the municipality.

9.1.1 The Better Way Forward

Sakon Nakhon is home to the sacred temple Phra That Choeng Chum, and the city has the honor of receiving support from King Bhumibol Adulyadej for the protection of Nong Han Lake. When the team of authors was first

Figure 9-1. (*Top*) Surface-flow constructed wetlands. (*Bottom, left to right*) Sunset behind the ponds; the town of Sakon Nakhon; a woman fishing at the wastewater treatment facility.

visiting the facility, the mayor of Sakon Nakhon, Mr. Komut, told us, "It was originally the idea of His Majesty the King to treat the wastewater of Sakon Nakhon in order to protect Nong Han Lake. Most people in Sakon Nakhon are glad that the treatment plant really works. We manage to treat wastewater from nearly all of the population of Sakon Nakhon." A nature-based solution with ponds and constructed wetlands was ideal in Sakon Nakhon because the municipality had vacant land available. Mr. Komut added, "One huge benefit to wastewater treatment with constructed wetlands is the low operating costs since the bill for electricity is much smaller compared to other treatment systems. Furthermore, it is a very efficient solution when you utilize natural processes to cleanse the water. It simply gives very impressive results." (Fig. 9-2).

To maintain those good results and to ensure that the public recognizes these achievements, Sakon Nakhon's wastewater treatment plant has worked hard to earn the International Standards Organization (ISO) 9000 quality certification. "I think we have one of the best treatment plants in the country and since it is based on the use of constructed wetlands, which is not a very

Figure 9-2. (*Left to right*) Municipal officer (Mr. Ruangrot), operations manager (Mr. Channachai), sanitation director (Mr. Somchat), and mayor (Mr. Komut).

well known technology, we are also creating a combined visitors and research center at the treatment plant," Mr. Komut added.

9.1.2 A Tour of the Constructed Wetlands

Mr. Channachai, plant operations manager, told us, "Basically, the wastewater is collected in pipe systems and channels located below the streets in Sakon Nakhon city center. Then the collection systems transport the wastewater and stormwater by gravity flow to two pumping stations." The wastewater treatment plant was located close to the town, minimizing costs for transporting wastewater for treatment. At the inlet, the wastewater is screened for solids, plastic, cloth, and other larger objects. The screen grid is automatically cleaned and collected solids are transported to the municipal landfill site. "After [the wastewater passes through] the automatic screen, four large electrical pumps lift the wastewater from the pump-sump to the distribution structure and from here it gravitates into the treatment ponds," Mr. Channachai said.

Pond processes. The first part of the municipal system consists of two parallel lines of waste stabilization ponds, and the wastewater is discharged into the ponds through a manifold pipe where the different openings can be closed to facilitate maintenance or sludge removal from the ponds Mr. Channachai explained, "In these ponds organic matter is primarily removed through algal photosynthesis and other physical and biological processes. The large size of the Sakon Nakhon wetlands means a long detention time, a lot of exposure to sunlight, and high temperatures helping to boost treatment efficiency." The wastewater is discharged into a collection channel from the pond system.

As a result of the algae production in the ponds, many of the pollutants in the wastewater are taken up by algae. According to Mr. Channachai, "The algae consist typically of green algae and blue-green algae, the latter being mildly poisonous to animals and humans. To reduce the concentration of algae in the discharge from the collection channel, we've introduced the screening of sunlight by floating plants, and other experiments will be implemented."

Constructed wetland processes. From the collection channel, the wastewater flows through weirs into six constructed wetland cells (Fig. 9-3). The wetland cells treat the wastewater for the final time before discharge into the lake. Remaining pollutants are, in other words, removed efficiently through various physical, chemical, and biological processes. Mr. Channachai went on, "The wetlands are constructed with varying depth and planted with a variation of plants, suitable for uptake of nutrients and removal of organic matter and bacteria." After final treatment in the constructed wetlands, the wastewater (now so clean that it easily meets the national standards for wastewater discharge) is directed into Nong Han Lake.

9.1.3 Thailand's First Municipal Plant to be Awarded ISO 9000 Certification

Requiring seemingly endless rounds of document preparation, ISO guidelines can sometimes seem tedious and overly time-consuming. But Mr. Ruangrot, an officer with the Sakon Nakhon municipality, thinks the hard work is worth the effort. "I'd recommend ISO to technical people at any treatment plant. It is an excellent way to set and meet quality standards," he enthused. "Many variations of ISO have been developed, but it is ISO 9000 that [was] chosen as the most appropriate and sustainable for the Sakon Nakhon treatment plant." In Sakon Nakhon, the ISO certification program was completed in just 10 months, in 2005. "The Sakon Nakhon wastewater treatment plant

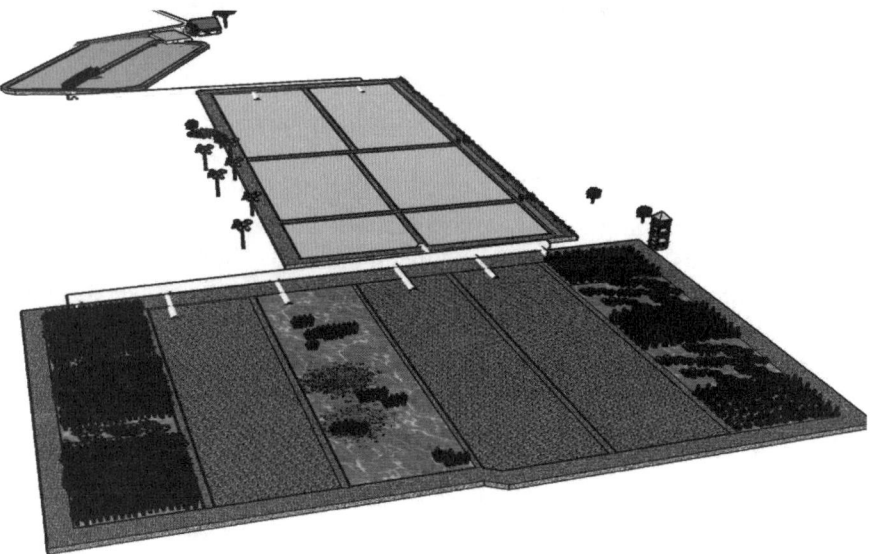

Figure 9-3. Layout of pond and surface-flow constructed wetland system.

is a ground-breaking facility in Thailand and certification will only help to cement its position at the vanguard of nature-based wastewater treatment," Mr. Ruangrot concluded.

By guaranteeing that key work processes are adhered to, ISO 9000 assures the water quality of the plant effluent. An ISO-certified plant agrees to perform key processes such as turning on pumps when required rather than having wastewater by-pass the plant; recording electricity consumption to monitor costs; having trained staff on duty at all times; and conducting regular water sampling.

Quality control is a key result of the ISO process—catching problems before they grow too large and providing ways to rectify them. Water quality sampling is done at regular intervals at a cost of about 5,000 baht per month ($150 USD). Because all work procedures are carefully examined for importance and then written down, the process of ISO documentation makes explicit the internal workings of the plant. It becomes a wastewater treatment "machine" that can be fine-tuned as needed.

Although the standards developed in this project pertain to wastewater treatment, the *process* of developing standards can be used in other municipality work. "Working on the ISO process has allowed us to see how to do good, well-planned work," said Mr. Ruangrot, "and we can apply those lessons to other municipality work." Having the nearby wastewater treatment plant operating under ISO 9000 rules assures local residents of the wastewater quality being discharged near their homes. Water quality standards are set and they must be maintained. If the standards start to slip, the treatment plant could lose its ISO certification, which could tarnish the image of the municipalities in charge; thus, ISO helps to maintain the political will necessary to maintain the treatment system. Mr. Ruangrot explained, "ISO procedures are written down. This allows for the continuation of the high standards even if personnel leave or the political situation changes. . . . The whole process is transferable."

Steps along the path to ISO 9000 certification include initial training on what ISO is; training on how to prepare ISO documentation; determining which documents are necessary to include in the final submission package; an internal audit to check the validity of regulations as spelled out in the ISO documents and to verify that the rules are being followed; and a third-party external audit that, if passed, allows for the awarding of ISO 9000 certification.

9.1.3 The Visitor and Research Center

Operating one of the country's most successful and ambitious wastewater treatment plants, it is only natural for the managers of Sakon Nakhon's facility to share their experiences with the public. "We want to create a visitors' center for the broader population," said the sanitation director, Mr. Somchat.

> **Box 9-1. Using International Standards**
>
> Driving past factories or commercial buildings, one often sees banners proudly declaring that that business has been certified as ISO 9000-compliant. In the nearly two decades since the first ISO 9000 standard was developed (based on an earlier British standard), the ISO process has permeated modern life in more ways than we realize. International Standard Book Numbers (ISBN), metric screws and bolts, and many, many other features of daily life are today standardized through the ISO. The International Organization for Standardization, or ISO, is based in Geneva, Switzerland. It is a worldwide federation of national standards bodies whose role is to promote the development of standardization. A standard like ISO 9000 defines the guidelines for an organization to follow to ensure that their clients' needs are met. The organization agrees to documentation that acts as a framework that regulates work processes, material procurement, training, and ways to continuously improve. The result is products and services that are of a guaranteed consistent quality that the customer can depend on. A wastewater treatment plant with ISO 9000 certification will run efficiently and effectively, constantly improving the management of the plant, and ensuring continued wastewater treatment. In Sakon Nakhon, ISO 9000 designation signifies international recognition of the viability of this constructed wetland technique, and the local people are assured that effluent from the plant meets national water quality standards.

Despite the fact that constructed wetlands are a very appropriate methodology, it is still not a well-known technology.

The visitors' center is oriented toward school classes, people from Sakon Nakhon with a particular interest in environmental issues, and tourists from Thailand or the rest of the world. "We do receive quite a few visiting school classes. Some days there are dozens of school kids at the treatment plant," Mr. Somchat remarked. "To improve the facilities and make the school class visits as valuable a learning experience as possible, the management has taken some concrete initiatives. Right now, we are implementing a beautification project which provides instructive signs with information on how the different elements of a treatment plant based on ponds and constructed wetlands work. We have also planted trees to make the whole area nicer and to protect the visitors from the sun in order to make the visit an enjoyable outing. In order to give our visitors a better overview of the whole treatment plant [the treatment plant covers about 70 hectares], we have also constructed a 7-meter-tall tower which serves as a perfect viewpoint from where you can overlook the entire area, including the Nong Han Lake [Fig. 9-4]. Combined with a signpost, this tower gives the visitor insights on the treatment plant

Figure 9-4. School group visit to the wastewater treatment plant.

as a whole." Three pedestrian bridges have also been constructed to improve accessibility to the plant for both staff and visitors.

"It is an obvious idea also to cooperate with professors and students from our two local universities in various sorts of field studies. They have already studied the retention of zinc, copper and lead," Mr. Somchat explained. Other areas where more research is needed include processes facilitated by the water hyacinths, the rock matrix interaction with algae, and the biological processes taking place in the sludge.

Research with local university students started in 2004. Sixty to 100 students have now studied the treatment plant, and it is not only the local university students who carry out field studies in the constructed wetlands. "We have also been working with professors and students from the universities of Khon Kaen and Chiang Mai. As interest continues to expand, we expect our research center to appeal to many researchers from throughout the country," Mr. Somchat said.

The wastewater treatment plant in Sakon Nakhon is a full-scale, living laboratory that treats the wastewater of some 50,000 people, improves the environment in and around Nong Han Lake, exemplifies the appropriateness of good management, and is the core object facilitating an innovative, interdependent, and self-perpetuating network of politicians, technicians, researchers, students, and professionals at large. The network ensures reliable and continuous operation of the plant as well as functioning as a catalyst for research activities and capacity development in a rural part of the country.

))) 9.2 Reflections on Appropriateness and Sustainability

The Sakhon Nakhon municipality wastewater management system clearly represents an appropriate and sustainable system (Fig. 9-5). We assessed its collection, treatment, and energy systems as being appropriate with a high level of simplicity, whereas others, like re-use and urban integration systems, are appropriate but could have been improved, especially during the design phase.

The *collection* system utilizes existing, well-functioning drainage systems and manages to efficiently get all wastewater to the treatment facility by gravity alone.

The wastewater *treatment plant* consists of a combined pond and constructed wetland system. Being one of the few such systems in the country, and having been operated effectively by the municipality for more than 15 years, clearly demonstrates sustainability.

The total treatment plant area is very large, covering some 70 hectares of reclaimed land that had once been a shallow part of Nong Han Lake. The municipal treatment facility was initially implemented as a pond-based system focusing on easy O&M and reasonable treatment efficiency. The treatment facility was later expanded with six surface-flow constructed wetland cells to improve treatment efficiency before discharge to the lake, taking the treatment capacity to 8,000 m^3 per day.

The combination of ponds followed by constructed wetlands is appropriate at treatment facilities where the wastewater flow has high organic matter content, which can cause clogging in gravel filter wetlands (horizontal- as well as vertical-flow subsurface wetlands). Or, high organic content can cause

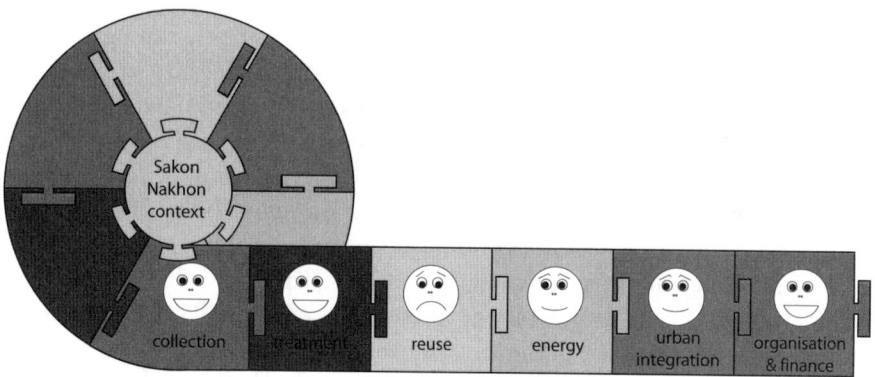

Figure 9-5. Contextual appropriateness scoring of the six elements of the wastewater management system at Sakon Nakhon.

Smile, contextually appropriate; no expression, somewhat appropriate; frown, not appropriate.

anaerobic conditions and die-off of plants in the inlet zone of surface-flow wetlands. By allowing initial treatment in the pond systems, the level of organic matter can be reduced before the wastewater is discharged to the constructed wetlands for further treatment. However, wastewater influent with high concentrations of organic matter has rarely been recorded at the Sakon Nakhon treatment plant; as such, both the significant scale of the plant and the addition of the constructed wetland part do not fully correspond to the scale of the problem. The pond system, especially of the size implemented, would suffice. Alternatively, a single inlet pond prior to the wetlands would probably be adequate to maintain the same treatment levels as today, while reducing the land use by about 30% to 40%.

Treated wastewater is discharged into the lake, so no *re-use* takes place. Because re-use would require pumping the wastewater, only re-use that would offset this extra cost could be justified. So far, the costs of tap water in the area (which is also sufficient during long dry periods) has not made pumping the treated water a viable option. This might come in future decades with higher prices for tap water; with possible longer rain-free periods as a consequence of climate changes; or if the costs of pumping water could be lowered with, for example, cheaper and more efficient solar-powered water pumps.

There are, however, two other side benefits. All the ponds in Sakon Nakhon are being used for aquaculture, and intensive fishing can be seen at all of the ponds on a regular basis. (Though the local fishermen wade into the ponds without protection, exposing themselves to pathogens in the wastewater.) Also, the frequent flow of motorcycles to the site proves that local florists and landscapers collect bulrushes and other plants in the wetlands, free of charge, indicating a significant demand for such plants for commercial purposes. The wastewater management system at Sakon Nakhon evidences the potentials for re-used water and sale of the byproducts of treatment by surface-flow constructed wetlands.

Energy is used to lift wastewater once, from the pumping station into the ponds via four large pumps. Thereafter, transport of wastewater to the treatment facility and distribution and flow in the ponds is entirely based on gravity. Furthermore, treatment is based on solar radiation and no additional energy supply is used. Thus, the pumps are the only energy-consuming elements in this wastewater management system and account for about 90% of total operating costs.

Urban integration is probably the weakest element in this system. The physical layout of the plant is very geometrical ("engineered"), inharmonious with the natural landscape of the area. The rational rectangular ponds contrast sharply with the softly curved lakeside, and the edges between the treatment facility and Nong Han Lake could have been softened with a much more environmentally sensitive physical design (Fig. 9-6). Instead of such strict, rational geometry, meandering curves, variations, surprises, unexpected meeting

Figure 9-6. Satellite image of Sakon Nakhon and the wastewater treatment plant consisting of rectangular ponds on the right.

Source: Google Earth, with permission.

places, ponds, and vistas—mirroring the characteristics and beauty of the local landscape—would have created a better fit to the surroundings and thus attracted more visitors. Instead of barbed wire and fences, distinction between the restricted and nonrestricted areas could have been established through the layout of the parks and canals.

The system was designed more than a decade ago, at a time no one paid attention to urban integration of wastewater management infrastructure. However, in the past few years the municipality has invested substantial funds to upgrade the treatment facility to make it more user-friendly, inviting, and interesting. Trees have been planted to provide shade and beautification, vistas and signage established, and bridges installed to improve access. The "engineered" feel of the treatment facility is slowly being modified and improved.

Organization and finance: If motivation and commitment exist at the political and administrative levels, the financial and technical capacity clearly exists in Sakhon Nakhon municipality to operate and maintain this simple wastewater management system.

This *if*, however, has been challenged a number of times during the life span of the system. During one period, political interest in maintaining the costs of operating the system waned and most of the wastewater by-passed the treatment facility for a while. During another multiyear period, intergovernmental disagreements disconnected the pond and constructed wetland system. The pond system was a municipality-designed and -managed project, whereas the constructed wetland system was designed and managed by the Fisheries Department, a central government institution. Due to a number of financial, legal, and land ownership issues, disagreements arose and for many years the systems were unlinked: wastewater was discharged right after the pond system treatment, resulting in the constructed wetlands receiving no wastewater in all that time. Only recently have those issues been resolved. Responsibility for the constructed wetlands has been transferred to the municipality, resulting in a recombined treatment system. This is another example of wastewater management systems being best off in the hands of the local municipality. For the past 5 years or so, management of the system has been local and efficient. The level of commitment and support from the mayor's office and the technical department has been good, and sufficient technical and managerial capacity has been developed at the wastewater treatment facility.

An appropriate technical system with a high level of simplicity provides the *basis* for sustainability, but certainly no guarantee. That guarantee often stems from whether the system "makes sense." The Sakhon Nakhon system barely passes that assessment. It makes sense to an environmentally friendly mayor, perhaps, but not to many other locals. The here-and-now benefits of treating wastewater at this end-of-pipe location for this huge lake are minimal and invisible, and only taking a long-term, environmentally conscious view would justify the present efforts and costs. The problem, again, is that the wastewater treatment primarily helps the environment but not the people in the area. In poor areas, this is normally a recipe for a failed wastewater management system.

))) 9.3 Smart Technologies at Sakon Nakhon

9.3.1 Combined Pond and Constructed Wetland System

9.3.1.1 The Technology: Combined Ponds and Free-Water Surface-Flow Constructed Wetlands

Technically, the treatment plant is designed as a combination of ponds and surface-flow constructed wetlands, which creates a vast grid of geometrically shaped lakes, shallow marshes, and stands of tall bulrushes divided by linear dikes and dusty red dirt roads. The design mirrors the cultivated landscape of rural northeastern Thailand (see Fig. 9-7 for an overall assessment).

Pollution reduction in ponds. The Sakhon Nakhon pond treatment system has been designed with six earthen facultative ponds—three ponds connected in two parallel series to increase operational efficiency. The first

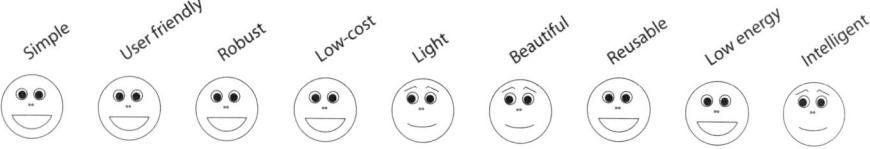

Figure 9-7. Pond and constructed wetland system: a smart technology?
Smile, supportive element for overall potential; no expression, somewhat supportive; frown, not supportive.

two ponds are large, 35,000 m² and 2 m deep, and the following two sets are smaller, 23,000 m² and 11,000 m², respectively, and only 1.5 m deep. All the ponds treat wastewater through an aerobic zone near the surface and an anaerobic zone near the bottom (Fig. 9-8). The ponds are large, manmade basins into which wastewater flows and where effluent is treated through the natural forces of sun, wind, gravity, and biological activity. It is a nonmechanical treatment process where water is transported through the pond system by gravity flow alone and water levels are controlled through simple weirs. The level of O&M is very low, almost limited to sludge removal every 20 to 30 years. Treatment in the pond systems is based on biological processes in which bacteria digest organic matter and nutrients and absorb it into their cells. The process employs algal–bacterial symbiotic interactions where algae provide oxygen through photosynthetic production, and bacteria degrade and use the organic matter for further bacterial growth. The dead organic matter has now been absorbed into a living organism, which later can be removed.

The pond system is well designed for the tropical climate of Sakhon Nakhon, where high temperatures and ample sunshine make it highly efficient. In addition, its low construction costs at (about one-tenth the cost of advanced activated sludge wastewater treatment plants in an area where land costs are not extraordinarily high) and very low maintenance requirements make it a robust and sustainable technology for Sakhon Nakhon.

Pollution reduction in surface-flow constructed wetlands. At a cursory glance, a surface-flow constructed wetland may look simple and inactive,

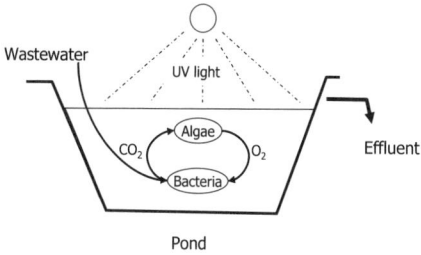

Figure 9-8. Pond treatment.

but in fact numerous biological, chemical, and physical processes are occurring simultaneously to remove contaminants from the wastewater that passes through it. In constructed wetlands, the plants, water, sand, and gravel create complex microenvironments where physical (e.g., sedimentation), chemical (e.g., adsorption), and biological (e.g., microbial decomposition) processes occur to remove pollutants from the water. The figure illustrates some key ways in which Sakon Nakhon's surface-flow constructed wetlands are cleaning the town's wastewater (Fig. 9-9).

In addition to the treatment processes taking place in the pond systems, the plants contribute to the removal of nutrients, suppression of algae, and sequestering trace organics. As the aquatic plants assimilate nutrients from the effluent, biomass is produced. To prevent secondary pollution from the micro-algae growth, or oxygen depletion and a release of phosphorous following plant die-off, plant biomass is removed from the system through harvesting the plant biomass.

Combined ponds and surface-flow constructed wetlands. The functions of the ponds are biodegradation of organics; removal of nutrients; nitrification/denitrification; ammonia volatilization; microbial uptake of nitrogen and phosphorous; and reduction of pathogenic microorganisms. The functions of constructed wetlands are removal of algal cells, filtration and sedimentation, further nutrient removal, nitrification/denitrification, and plant uptake. Combined, these systems provide very effective treatment of wastewater and offer potential benefits through water re-use and recycling the organic byproducts. The use of oxygen produced by growing plants can eliminate

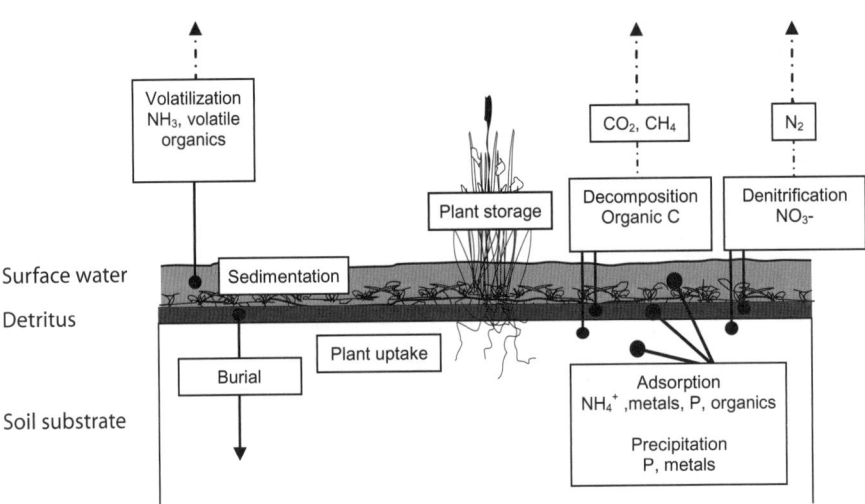

Figure 9-9. Surface-flow constructed wetland treatment.

the cost of aeration in designs or settings where aeration is needed, especially in warm countries with a year-round growth season, by utilizing symbiotic relationships between heterotrophic bacteria producing carbon dioxide and micro-algae producing oxygen.

In addition, these systems entail low construction and O&M costs. Neither the ponds nor the constructed wetlands require advanced technologies during construction—only simple construction works such as earthworks, inlet and outlet channels, erosion protection works, hydraulic concrete structures such as overflow weirs and V-notches, access and service roads, service buildings, and initial plantings. There is little reliance on experts to maintain complex machinery. Instead, maintenance focuses on keeping waterways clear.

Advantages such as cheaper operation and simpler maintenance are meaningless if the constructed wetland systems do not effectively remove contaminants from wastewater. The pond and constructed wetland system in Sakon Nakhon has been in operation for more than 10 years, and water sampling proves the system works. For the entire time of operation the municipal wastewater department has been monitoring the efficiency of the treatment plant through analysis of key pollutants; these tests have shown that wastewater treatment in the system have continuously resulted in 60% to 90% elimination rates for BOD, nitrate, phosphorous, and suspended solids.

9.3.1.2 Technical Considerations, Balances, and Choices

Sizing. Designing the size of a combined pond and wetland system is difficult because many processes interact, making it hard to make exact predictions or determine precise design criteria. The Sakon Nakhon system was designed more than a decade ago and still works efficiently, partly, as mentioned, because it was overdimensioned. For design purposes, wastewater volumes and quality might be known and population-equivalent (PE) guidelines for sizing of ponds and constructed wetlands exist, but to make the sizing of combined systems more accurate, more knowledge and more practical case studies upon which more accurate generalizations can be made are needed.

Choice of the pond system. Aggravating the difficulties of sizing combined systems are the many technical options for each system. Options regarding waste stabilization ponds include, for example, anaerobic ponds, aerobic ponds, facultative ponds, aerated ponds, and fish ponds. In Sakhon Nakhon, only facultative ponds were constructed. Alternative options for the Sakhon Nakhon system could include, for example, (1) an additional maturation pond to reduce the amount of bacteria before re-use or discharge. This pond could double as a pond for fish farming; (2) a rock matrix to reduce the concentration of algae prior to discharge to the surface wetlands; (3) a facultative pond designed with a pit hole about 6 m deep to promote sedimentation of wastewater solids and anaerobic decomposition of methane; or (4) an additional 5-m-deep anaerobic pond for organic removal by sedimentation of solids and anaerobic digestion in the resulting sludge.

Odor control. Surface water in wastewater ponds or constructed wetlands has a risk of odor problems. This is especially true if anaerobic conditions occur without good management practices that either collect biogas from anaerobic pits in facultative ponds, or cover the surface with duckweed or other floating plants to reduce fermentation. At Sakon Nakhon, odor problems are not an issue because the system is far from permanent residences; winds from the lake reduce occasional odor concentrations; and the ponds and wetlands were designed for aerobic conditions and generally do not experience anaerobic conditions due to low concentrations of organic matter in the influent.

Mosquito control. Another issue of concern in combined systems is the risk of mosquitoes breeding in the still water. Although in general mosquitoes prefer clean water rather than polluted aquatic environments, the proportion of still or open water surfaces should be reduced. Other controls include using windy locations such as lakesides; cutting grass and other vegetation along the slopes of the ponds; promoting aquatic polyculture, including fish species likely to eat mosquito larvae; or (in anaerobic ponds) allowing duckweed to form a complete mat over the water surface to prevent mosquito larvae populations from reaching the surface. Water hyacinths are known to promote mosquito breeding, although systematically cropping them and clearing the pond of dead plants and decaying plant matter can reduce the number of larvae. At Sakon Nakhon the large surfaces of the ponds allow wind to create waves and turbulence on the water surface; this has been so effective that no major mosquito problems have been identified.

Fishing and public access. Although never planned for or officially approved by the municipality, all of the ponds in Sakon Nakhon are being used for aquaculture and people fish the ponds on a regular basis. Allowing fishing in the ponds opens up the treatment system to the public and provides income for poor residents in the municipality, as well as being a recreational activity. However, there are concerns about public health risks due to the potential pollution in fish that is used for human consumption, and reducing the amount of fish in the ponds, which play an importing supporting role in the treatment of the wastewater, thereby negatively impacting treatment efficiency. Fishing in the ponds is too intensive to allow the fish to grow to any reasonable size. Introducing water hyacinths could allow the fish to hide under the floating plants, thereby enhancing fish breeding conditions inside the entire facility. Another approach would be to prohibit fishing altogether.

9.3.1.3 General Reflections and Wider Considerations

Combined pond and constructed wetland systems have only recently been implemented in developing countries, but they should be more widely used because they are an effective and sustainable wastewater management method. They can effectively treat low-BOD wastewater year-round at a fraction of the cost of conventional mechanized systems. By combining different types of ponds and constructed wetlands, treatment systems can be designed to accommodate a wide variety of polluted waters, including domestic wastewater, industrial wastewater, and stormwater run-off.

The major advantages of combined pond and surface-flow constructed wetland systems are robustness, cost-efficiency, and flexibility in the sense that the technology can be adapted to future needs—it is amenable to rehabilitation or changing layouts and land use; it is not locked into a single technological design solution.

10

Wastewater Planning in Pathumthani Province: Appropriate Planning of Large-Scale Wastewater Management

⟩⟩⟩ 10.1 Thinking Small, Big Scale

Many examples exist of best practices of appropriate and sustainable wastewater management at an on-site or cluster scale. Fewer examples of large-scale applications exist. The Pathumthani Province case provides a feasibility study for applying appropriate on-site and cluster wastewater management systems on a large scale in developing countries.

Pathumthani is located directly north of Bangkok and is part of the Bangkok metropolis. The province is located in the low alluvial flats of the Chao Phraya River, which also flows through the capital city. Many canals cross the province, drain the area, and feed the local rice paddies. The province has a total population of approximately 500,000 people and 11 catchments covering 95 km². Based on an earlier feasibility study, a centralized wastewater management system was planned to be installed to service this area. That study provided a very expensive solution, and some in the government also doubted its effectiveness. The Wastewater Management Authority of Thailand authorized a new study, which is described here, to assess three alternative on-site or clustered system approaches and compare them with the previous study that had led to the centralized option.

Many professionals consider on-site and clustered pond and wetland treatment systems the most feasible and appropriate wastewater management option for smaller urban and rural areas. In addition, it is commonly agreed that in dense downtown metropolitan areas, where land is scarce and expensive, technologies requiring less land are better solutions. But what about in large, "semi-dense" suburban areas? Should these areas implement the "downtown" or the "provincial" wastewater management option, or are other technical and financial options available for these suburban areas? These were the key questions the new feasibility study addressed. Another issue was the methodologies available to plan for large-scale wastewater management systems in urban areas in developing countries.

The answers to these technical and methodological questions are elaborated below, in a step-by-step description of the alternative feasibility study. Hopefully this can inspire professionals to examine and perhaps implement more large-scale, decentralized wastewater management systems and to continue to share and improve the methodologies of such planning exercises.

The new Pathumthani feasibility study contained 8 components, which were divided into 19 analytical steps.

10.1.1 Component 1. Determining Feasibility Options

Step 1. Define Typical Characteristics

To identify suitable future options, we assessed the existing general wastewater management issues in Thailand. Our alternative Pathumthani analysis considered various important fundamentals (see also Chapter 2), and found that four main features characterized wastewater management in the country:

1. Sewer/drainage systems have been built in almost all cities and almost all sewer systems separate sewage into greywater (bathing, laundry, etc.) and black (toilet) wastewater.
2. In most cases only part of the wastewater reaches established central treatment plants due to overflows in the sewers caused by sediment blockages, lack of cleaning, broken pipes, and malfunctioning pumping stations.
3. Existing treatment plants are often oversized and/or unnecessarily advanced compared to inflow and effluent standards.
4. Municipalities often have financial and technical difficulties in operating and maintaining the mechanical elements in the pumping stations and at the treatment plants.

Step 2. Define Guiding Principles

Based on the above assessments, we chose the following nine general principles to inform the feasibility options we would select for wastewater management in the Pathumthani project area:

1. Maintain the principle of separation of black and grey wastewater.
2. Collect and dispose of black wastewater.
3. Improve the performance of the greywater collection system.

4. Use appropriate and sustainable wastewater collection technology.
5. Ensure basic but sustainable treatment of greywater.
6. Use appropriate and sustainable treatment technologies.
7. Separate the treatment systems for domestic and industrial wastewater.
8. Implement local re-use or discharge of treated wastewater.
9. Implement effective sludge handling.

10.1.2 Component 2. Assessment of Project Area Characteristics Relevant to the Feasibility Study Options

Step 3. Analyze and Define the Site Characteristics

To identify the best possible future options, we assessed the specific characteristics of the project area. The total Pathumthani project area consisted of 11 catchments covering 95 km^2 and included nine political units, four municipalities and five local administrations (Fig. 10-1 left).

We found that five main features characterized the existing handling of wastewater in the project area, and that these strongly influenced existing and future wastewater management options in the project area:

1. *Flat topography.* The area was topographically very flat, with spot levels ranging from 1.5 to 3 m. The groundwater level was 0.5 to 1 m below the surface due to poor soil permeability (river sediments dominated by clay). Groundwater in the area was generally not extracted for water supply intake. These aspects were important in relation to construction of sewer lines and identifying potential locations of treatment plants.
2. *Eleven catchment areas.* We identified 11 overall catchments (Fig. 10-1 right). Within these were a number of subcatchments (existing main discharge points). The identification of subcatchments was important for determining the locations of interceptors and treatment plants. Catchment 4 (the pilot area in the new study), for example, contained five subcatchments (refer to Section 10.1.4, Component 4).

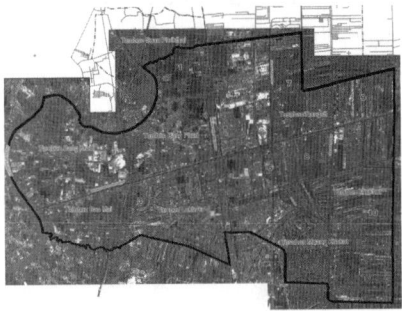
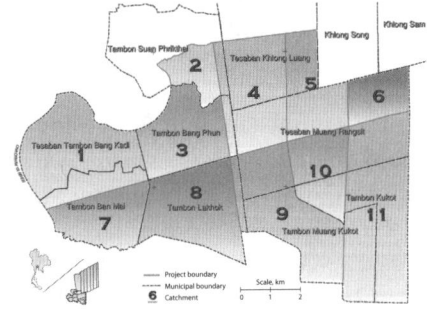

Figure 10-1. Location and catchments in the prefeasibility study area.

3. *Highly uneven land development rates.* The catchment areas within the project area varied highly in terms of development. Some catchments had urban land development of 20% whereas others were 70% urbanized. Different areas also had strikingly different expected future development rates. These aspects were important for determining appropriate and sustainable collection and treatment systems for the different areas.
4. *Uneven suburban structure.* Most of the urban areas were dominated by two- and three-story buildings (combined shops and residences) along the main roads, several markets, and single- or two-story domestic dwellings inside the housing areas along minor roads and alleys. Some of the urban areas were very densely populated because they contained numerous three-floor housing blocks with single-room apartments for workers. Along channels and rivers were a number of squatter homes on stilts. The urban areas also included institutions such as schools and government buildings as well as small-scale enterprises such as restaurants, garages, and shops. There were some large-scale industrial complexes within each catchment as well as one very large shopping complex. This distribution was important for calculating different wastewater loads from different types of urban structures.
5. *Existing wastewater management.* Almost all the buildings in the area were served by individual septic tanks, mostly constructed as seepage pits with gravel filters. De-sludging of individual septic tanks was done only when septic tanks were blocked. Some apartment buildings had individual treatment facilities. Squatter homes had latrines discharging to the nearest water body or drain. Overflow from septic tanks, greywater, and stormwater were discharged to rivers at natural or constructed discharge points.

Step 4. Define the Study Approach

The new overall study approach included (1) considering and selecting the most appropriate technology based on area characteristics; (2) assessing a pilot area containing typical land use types; and (3) estimating sustainable solutions for the full project area in terms of wastewater production, technology, and costs, and comparing these with the previous centralized wastewater management study that had been conducted for the same area.

Due to limited time and resources for this feasibility study, we selected a pilot area of about 10.2 km^2 for detailed field survey and data analysis. Based on this and other preliminary surveys, and data collection for the total project area, we calculated projected estimates for the full 95-km^2 project area.

We assessed the project area in relation to existing and future conditions based on three factors: land use, wastewater production, and wastewater collection and treatment systems. Existing land use was identified using satellite images. In the pilot area the existing wastewater production was based on comparison of population-equivalent (PE) calculations with actual wastewater discharged through main outlets, combined with analysis of subcatchment service areas. In the project area the existing wastewater production was based on PE calculations (based on the generic data derived from the pilot area) and service areas. Projections were extrapolated by spatial analysis of predicted urban land use development.

Our data were primarily self-generated, utilizing up-to-date primary input based on satellite images, ground check surveys, wastewater flow measurements and sampling at key discharge points, and classifications (land use, wastewater production, etc.) for generalization purposes. Our analysis did not rely on secondary data.

We purchased satellite imagery of the project area and produced geographic information system (GIS) maps. Our key maps showed the project and pilot areas; other maps of the areas showing catchments and discharge infrastructure were based on satellite imagery and technical background maps; PE maps showed population density and calculated loads; land use maps showed the existing land use as analyzed from satellite images and ground check inspections; and project proposal maps showed the study areas and the proposed construction works.

10.1.3 Component 3. Define Appropriate and Sustainable Technology Options

Step 5. Select Wastewater Management Options
Many different collection and treatment options could have been applied to the project area. We chose three options relevant to the typical wastewater management issues outlined in Step 1. For the characteristics of the project area defined in Step 2, we selected (Fig. 10-2):

- *Catchment wetland:* Wastewater intercepted at catchment outlets and pumped into a constructed wetland treatment plant.
- *Subcatchment treatment:* Wastewater intercepted by gravity and treated in cluster treatment plants using constructed wetlands or similar systems.
- *Mini-treatment:* Wastewater treated in activated sludge mini-treatment plants connected to existing sewer lines under or next to roads.

Step 6. Define and Outline Each Option
Option 1: Catchment Wetland. This option principally included the establishment of a catchment interceptor sewer line intercepting wastewater from the main outlets within the catchment, and transporting the wastewater by gravity or pressure pipe to one treatment location within the catchment. At this

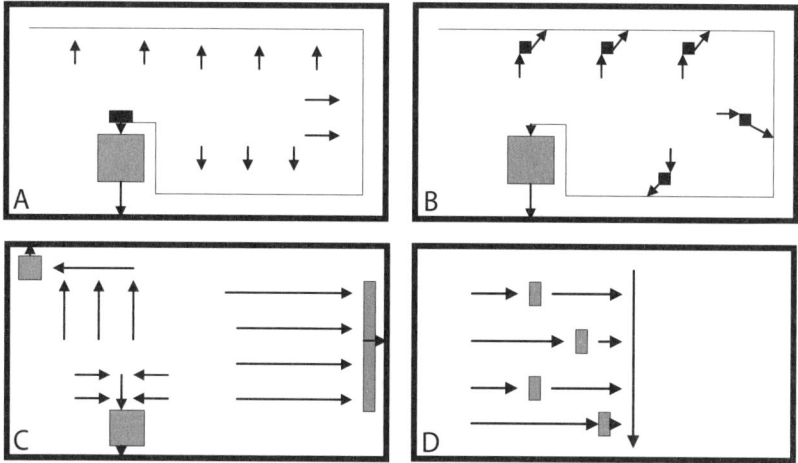

Figure 10-2. Principles of wastewater collection. (A) The gravity and pump principle. (B) The pressure-pipe principle (Option 1). (C) The subcatchment principle (Option 2). (D) The mini-treatment principle (Option 3).

Arrow line, wastewater flow in pipe; *grey box*, treatment facility; *black box*, pump.

location the wastewater would be directed or pumped into the constructed wetland treatment facility. The wastewater would flow by gravity through the facility and discharge by gravity into the nearest waterway or onto land for re-use purposes. In this option, wastewater would be treated either by a pond/surface-flow constructed wetland system or by a vertical subsurface-flow constructed wetland system. We estimated the cost elements for each of these techniques.

To optimize wastewater collection and minimize the need for pumping stations and transporting over long distances, we performed an overall catchment analysis. This analysis determined catchment boundaries such as rivers, major canals, railway tracks, and similar obstacles to gravity flow. The catchment analysis also included an assessment of open areas suitable for the location of a treatment facility.

Option 2: Subcatchment Treatment. This option principally included the redirection of wastewater from a sewer/drainage line and then by gravity flow to a local treatment location. Treatment would take place during gravity flow through a vertical-flow constructed wetland treatment system or use of other land-based cluster treatment systems such as existing low-lying wetland areas or aerated treatment in existing lakes. Wastewater would be discharged by gravity, or pumped during high tide, into the nearest waterway or onto land for re-use purposes.

The key components in the wastewater collection system for this option consisted only of connection piping and overflow structures—piping to redirect dry-weather wastewater into the treatment facility. The length would depend on the distance of the treatment facility from the existing pipes. Overflow structures for stormwater would be needed at the location where wastewater would be redirected.

Wastewater treatment would take place using one of the wetland treatment types described under Option 1. Because this option was linked to subcatchment and gravity interception, availability of land became more of an issue (in Option 1 wastewater would be pumped, resulting in more flexibility regarding choice of locations). Because vertical-flow constructed wetland treatment requires less land, this technology would have been more appropriate for Option 2.

We performed a subcatchment analysis that focused on existing flow directions, gravity systems, and main outlets. The subcatchment analysis evaluated drainage/sewer systems, main outlets (for the pilot area, also the quantity and quality of wastewater at the main outlets), and open areas suitable for the location of a treatment facility. Because the subcatchments were quite different, implementing a subcatchment approach would have resulted in substantially different assessments and proposed solutions for the various subcatchments.

Option 3: Mini-Treatment. This option included the redirection of wastewater from a sewer/drainage line into a mini-treatment plant located on or adjacent to the sewer/drainage line. Treatment would take place in a number of underground mini-treatment units (Fig. 10-3) located throughout the area at points where sufficient amounts of wastewater with sufficient BOD loads had been generated. After treatment, wastewater would be discharged back into the existing sewer/drainage system.

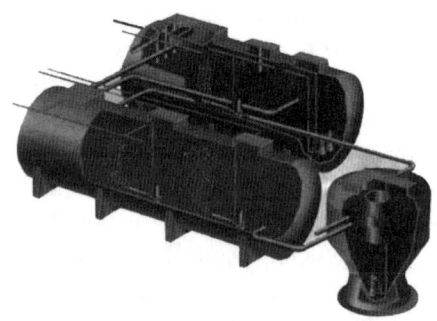

Figure 10-3. The mini-treatment system, measuring approximately 3.5 m × 6 m × 4 m.

The treatment plants would be connected to the existing sewer/drainage pipes. The key components in the wastewater collection system were therefore only connection piping and overflow structures—piping to connect dry-weather wastewater with the treatment facility, and overflow structures for stormwater. The surface level of the wastewater in the pipe would be very close to the surface level of the receiving water, and the water level in the pipes would remain unchanged upstream of the treatment plants. An overflow/by-pass would ensure that the inflow to the treatment plants would not exceed the capacity of the plants, especially during rain. The treatment facility would consist of an activated sludge mini-treatment plant, which could be designed to serve from 200 to 3,000 PE. However, in this study only the largest capacity was applied. The treatment system consists of two components: a combined equalization, aeration, and aerobic tank, and a sedimentation tank. After passing through a primary screen, the wastewater would be released into the equalization tank where suspended solids would be removed. The wastewater would be pumped into a flow control box to maintain its flow and density while being aerated in the aeration tank. The sedimentation tank would be used to settle sediments, most of which would be returned to the aeration tank as sludge, while excess sludge would be aerobically accumulated within the system.

The land requirement for a mini-treatment facility with a capacity of 600 m^3 per day would be about 275 m^2 (or 9 m × 32 m), which amounts to about 0.09 m^2 per PE. The width, length, and height of each facility was estimated to be 3.5 m × 6 m × 4 m.

10.1.4 Component 4. Scaling Down before Scaling Up: Detailed Site Analysis of One Catchment

The pilot catchment area covered 10.2 km^2 and contained the most densely built-up catchment area in the entire project area (Fig. 10-4 left). The area had a mix of different land uses but also an existing high density of residences, and was therefore a difficult location in terms of alternative wastewater management planning.

Step 7. Determine Present Land Use

Analysis of the satellite images produced an outline of the existing land use distribution. Residential land use was 30%, commercial and institutional was 10%, infrastructure was 5%, and 34% of all the land was open (Fig. 10-4 right). These areas, however, contained large closed-down governmental enterprises expected to be converted into residential land use.

The densely populated areas were mainly located in the southern part of the catchment, whereas industrial areas and new developing residential areas were located in the northeastern part. A very large shopping center, Future Park Rangsit, was located in the southeastern corner.

Figure 10-4. Pilot area: satellite image and land use analysis (2004).

Source: QuickBird Satellite Imagery/CD-WMA (Capacity Development for the Wastewater Management Authority, Thailand), 2004.

Step 8. Determine Present Population and Population-Equivalent

To estimate the population in the pilot area, we did a house count using the satellite images and then cross-checked it through a detailed ground survey of the types of land use and number of people in the different categories. We used commercial, institutional, and industrial factors based on national experience to estimate the population-equivalent (PE).

The resulting total PE for the area was 99,728: 76,823 in residential areas, 19,583 in commercial and institutional areas, and 3,322 in industrial areas. A 3-D spatial distribution of the PE figures indicated an uneven distribution of PEs, with most of the PEs found in the southern part and in one area at the center of the catchment (Figs. 10-5 and 10-6).

Step 9. Determine Present Wastewater Management Infrastructure

The existing infrastructure for wastewater management in the area encompassed the separation of black and grey wastewater for almost all units (residential, commercial, and institutional). Exceptions were mainly settlements located along rivers or on lake banks, but these were few in number.

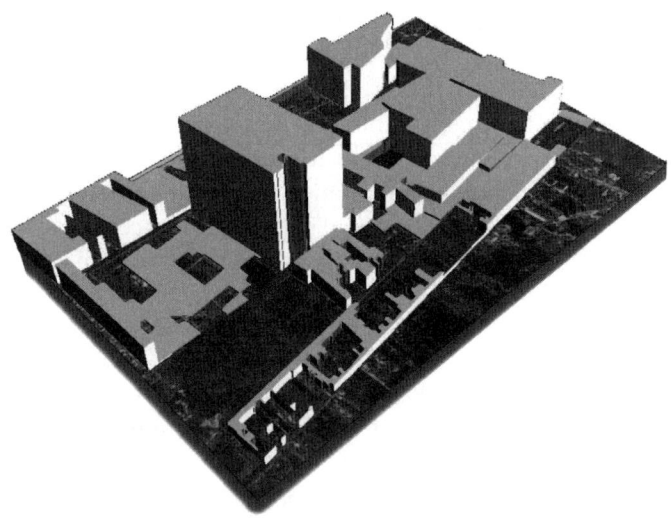

Figure 10-5. PE distribution in the pilot area seen from the northwest (2004). PE is shown as the height of each zone.

Blackwater was being led to and treated in individual septic tanks (with varying treatment effectiveness, mainly depending on the age of the septic tanks and the frequency of emptying). A good collection system for greywater existed in most of the subcatchments within the area, covering 86% of the PE.

We estimated the number of already constructed and established septic tank treatment systems to be about 8,400. This gave an estimated total cost of already implemented investments of about 150 million baht ($4.5 million USD) for septic tank treatment in the pilot area alone.

Some of the septic tank systems had been established more than 20 years ago and were expected to be less efficient, whereas tanks established in other areas were newer and were expected to be more efficient. An indication of this difference in septic tank efficiency was indicated in the higher BOD levels registered in the study at Outlet 1 covering the old city zone 11, compared to the BOD levels at Outlet 2 covering mainly the new city zones 6 and 12 (Fig. 10-6 right).

Almost all the greywater, combined with stormwater, was collected in a well-constructed and closed drainage system, and was discharged from the area through three main and two small discharge points—all discharging south into the Rangsit River. Greywater was, in general, collected and transported under hygienically acceptable conditions. Only a few open drains were found.

Many zones, including zones 1, 2, 4, 5, 9, and 10, were not connected to the sewer/drainage system and they discharged either through seepage or through open canals to open areas, mainly low-lying wetland-type areas. Two lakes also served as discharge areas, mainly for stormwater and slum settlements on the lake banks. These areas, however, accounted for only 14,100 PE or 14% of the PE in the total area.

No central treatment took place for black or grey wastewater, but at least 22 on-site wastewater treatment plants had been constructed and were, in varying degrees, in operation for major industries and commercial and institutional complexes, including the Future Park Rangsit shopping complex. We estimated that the total cost of already implemented investments for commercial or industrial on-site treatment in the pilot area alone was about 110 million baht ($3.3 million USD). Because access was not permitted to all treatment systems in the pilot area, the actual total investment was expected to be higher.

Step 10. Determine Present Wastewater Production

We assessed wastewater production using two different methods.

Actual Volume at Discharge Points. We estimated wastewater production based on actual volume through 24-hour measurements at the three main discharge points in the pilot area. The flow and BOD rates for each of the points are provided in Table 10-1. Almost all wastewater produced *and* discharged to drainage pipes in the area was discharged through the three main outlets analyzed (the three rightmost discharge points identified in Fig. 10-6).

Figure 10-6. PE per type, zone, and wastewater catchment areas (sewer and flow system, main outlets, and connected and nonconnected areas) (2004).

Table 10-1. Wastewater Production Based on Discharge Measurements

Outlet No.	BOD	COD	SS	Wastewater (m³/day)
1	146.5	201.9	54.8	1,220
2	55.6	111.1	39.7	15,154
3	33.4	79.1	20.6	2,923
Total				19,297

BOD, biochemical oxygen demand; COD, chemical oxygen demand; SS, suspended solids.

We noted that outlet pollution concentrations for Outlets 2 and 3, transporting about 94% of all the wastewater discharged from the catchment, were very low with BOD levels from 33 to 55 mg/L. We also noted that Outlet 2 alone discharged about 80% of all the wastewater from the catchment.

Analysis of the discharge frequencies during the registered 24-hour period revealed that Outlet 1 had a normal fluctuating discharge pattern for domestic wastewater and Outlet 3 had a constant discharge pattern, which might be attributed to the outlet mainly receiving treated wastewater from the large shopping complex. Outlet 2, however, had a constant, high discharge pattern of about 15,000 to 17,000 m³ per day, indicating either a very high infiltration rate at about 7,000 to 10,000 m³ per day or, less likely, 24-hour constant industrial discharges. If this proved correct, rehabilitation efforts could correct the infiltration rate, which would significantly influence the design criteria and cost calculations in the study. For example, instead of needing to collect and treat 19,000 m³ per day, sewer repair work on Collection Line 2 might result in the need to collect and treat only 12,000 m³ per day.

Using this method, we estimated the total wastewater production in the area discharged through the three main outlets to be about 19,297 m³ per day.

Wastewater Production Estimation Based on the PE Estimations. We used the following formula to estimate water and wastewater production:
- Average water consumption was set at 250 L per day per PE.
- Of the water consumed, we estimated that an average of 80% was discharged as wastewater.
- Of this figure, we estimated that an average of 80% was greywater being discharged into the sewer/drainage system, while an average of 20% was black wastewater discharged into septic tanks.
- We assumed an average infiltration rate of 20%. See Box 10-1 for a methodological discussion of these assumptions.

Given these assumptions, we calculated that the total wastewater production in the pilot area was about 19,148 m³ per day (Table 10-2).

Based on the above wastewater production estimations, we estimated the existing wastewater production per residential/commercial/industrial discharge

Table 10-2. Wastewater Production Based on Population Equivalent Calculations

	Population Equivalent (PE)[a]	Average Water Consumption @ 250 L/day/PE (m³/day)	80% Average Wastewater Production (m³/day)	20% Black (m³/day)	80% Grey (m³/day)	Adjusted for 20% Water Infiltration Rate (m³/day)
Residential	76,823	19,206	15,365	3,073	12,292	14,750
Commercial	19,583	4,896	3,917	783	3,133	3,760
Industrial	3,322	831	664	133	532	638
Total	99,728	24,932	19,946	3,989	15,956	19,148

[a]PE = 250 L/day

> ### ▶▶▶ Box 10-1. The Population-Equivalent Methodology
>
> Because we used the PE method for calculating wastewater production in the full project area, we compared this with actual flow to verify the accuracy of the assumptions made in the PE method. Based on the actual figures, the PE methodology was adjusted because the water consumption estimate was raised from 220 to 250 L/day to get a better fit between the two figures and thereby be able to use the PE methodology throughout the project area.
>
> It should be noted that one of the key methodological lessons learned in our study is the danger of using the PE method as a basis for wastewater management planning in developing countries:
>
> - The *actual flow measurements* showed the need to adjust the PE assumptions to get a better fit.
> - The *analysis of daily fluctuations per outlet* showed the need to adjust for possible high infiltration rates in certain sewer lines.
> - The *spatial analysis* showed the need to adjust the PE for the actual connected areas.
> - The *analysis of subcatchment flows and connections* showed the need to adjust the PE downward accordingly.
>
> The number of assumptions needed to use the PE equation introduces a very high level of uncertainty:
>
> - *Assumption 1: PE is known.* The actual PE is highly uncertain. For domestic PE in a developing country, it is hard to determine how many people actually live in an area. In our survey we estimated that the registered population was wrong by a factor of 2 to 4. For example, the Kukot district (Catchment 6) has a registered population of about 20,000 but our field interviews indicated that 40,000 to 50,000 people live in the district. The satellite house count suggested a population of 49,624. Conversion of commercial and industrial consumptions to PE figures is an even more uncertain exercise.

> - *Assumption 2: All PEs are connected.* The spatial and ground surveys showed that actual connection rates were highly variable—ranging from 14% not connected in Catchment 4 (the pilot area) to 100% not connected in Catchment 7.
> - *Assumption 3: Water consumption is known.* It was not possible for our study team, or for the earlier centralized feasibility study project, to get data on actual water consumption. This figure therefore had to be based on national rules-of-thumb consumption rates.
> - *Assumptions 4, 5, and 6: The water-to-wastewater and greywater-to-black wastewater and infiltration ratios are known.* These ratios had to be based on national rules-of-thumb ratios. Especially for the infiltration ratio, our analysis in the pilot area indicated that this ratio in certain circumstances might be highly inaccurate.
>
> We concluded that the PE method should only be used as a backup and that in future wastewater management planning the main method should include:
> - Actual wastewater flow and quality measurements for each main outlet in the catchment study area
> - Analysis of daily fluctuations per outlet
> - A spatial satellite imagery-based analysis
> - An analysis of subcatchment flows and connections.

category from each of the 14 zones within the catchment. We used these figures for the detailed planning of collection and treatment systems within the catchment.

Step 11. Predict Future Wastewater Production

We conducted a spatial analysis of future land use which, together with interviews with the local authorities, resulted in a projected distribution of land use in 25 years (Fig. 10-7 right). The prediction was that residential land use would increase from 30% to 54%, commercial and institutional would increase from 10% to 15%, industrial would decrease from 21% to 16%, open areas would decrease from 34% to 10%, and infrastructure land use would remain at 5%.

Based on this 25-year projection of land use distribution, and based on the combined actual flow and PE wastewater calculation method described above, we estimated that wastewater production in the pilot area in 25 years would be about 33,070 m^3 per day (Fig. 10-7 left).

10.1.5 Component 5. Detailed Application of Alternative Systems for the Pilot Catchment

Step 12. Analysis of Catchment Applications for Each Option

The PE-based wastewater production was calculated to be 19,148 m^3, rising to 33,070 m^3 in 25 years, serving a PE of about 150,000. However, planning

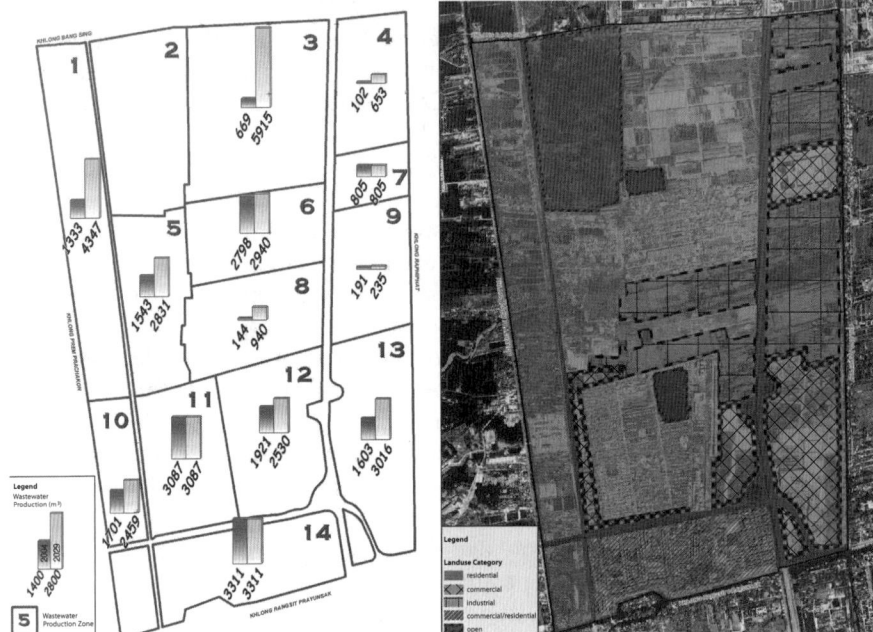

Figure 10-7. Projected (2029) wastewater production per zone and land use distribution.

with this figure had to be adjusted for development rates and service areas. A number of complicated, context-specific assumptions were made regarding development in the 14 zones in the area. Based on these assumptions, we calculated the expected wastewater production in the different zones and overall. We assumed the BOD level of the wastewater would remain in the same range; we used BOD 70 mg/L for design purposes.

Option 1: Catchment Wetland. This option implied one management plan for the three local districts covered by the pilot catchment area (indicating a technical solution with political complications). The total catchment area was covered by three main and two minor wastewater collection systems discharging into Rangsit River through three main and two minor outlets. These collection systems would be utilized as-is.

We proposed construction of an interceptor covering wastewater collection systems 1, 2, 4, and 5. At each outlet the interceptor system would pressure-pump the wastewater to the proposed treatment plant located in the upper part of the catchment, and then direct the wastewater by gravity flow through this facility before discharge into the western drain/river (Fig. 10-8).

Collection system 3 covered zone 13 with a PE-estimated wastewater production of about 1,603 m³ per day. For this collection system, no additional

Figure 10-8. Option 1 applied to the pilot area (*left*) and close-up of the location of the proposed constructed wetland (*right*).

collection or treatment system was proposed. The existing collection system mainly covered the large Future Park shopping complex. The actual measured discharge volume was 2,923 m³ per day with a BOD of 33.4 mg/L. This low BOD indicated that the already established treatment facility was functioning, but insufficiently. We decided that the most appropriate solution would therefore be to increase the efficiency of the existing treatment facility and/or increase enforcement activities. The collection system would thus cover four main outlets with an estimated wastewater collection of 30,053 m³ per day by the year 2029.

Because the discharge from the area had a low BOD level, we considered treatment by ponds and constructed wetlands an appropriate and sustainable treatment technology. We produced two different wetland designs and cost calculations for comparison—one using a pond/surface-flow constructed wetland system, the other using a vertical-flow constructed wetland system. Total investment costs, including the pressure-based collection system, for the pond/surface-flow wetland option, were calculated to about 520 million baht ($15.6 million USD) with an O&M costs at 0.26 baht ($0.008 USD) per m³ treated wastewater, compared to 470 million baht ($14.1 million USD) and

0.84 baht ($0.025 USD) per m³ treated wastewater for the vertical subsurface-flow wetland option.

Option 2: Cluster Treatment. This option involved the redirection of wastewater by gravity flow into three different treatment systems (Fig. 10-9).

- *System Center—Lake:* For the central part of the subcatchment (collection system for Outlet 2), the proposal was to utilize the existing collection system but to redirect the wastewater downstream of zone 8 into an aerated treatment system at the lake in zone 12. The collection system thus would cover one main outlet with an estimated average wastewater collection of 4,920 m³ per day in the year 2029 (zones 6, 7, 8, and 9). The treatment system for the center part of the cluster treatment system would be an aerated treatment system at a private lake in zone 12; its specifications would include lake improvements; 3 rai (4,899 m²) lake surface area; inlet and outlet structures; aeration equipment, aeration control; and a subsurface partition curtain.
- *System North—Wetland:* For the northern area, the proposal was to construct an interceptor system to cover zones 3 and 4, to be discharged into a wetland at zone 2. The collection system would thus cover one main outlet with an estimated average wastewater collection

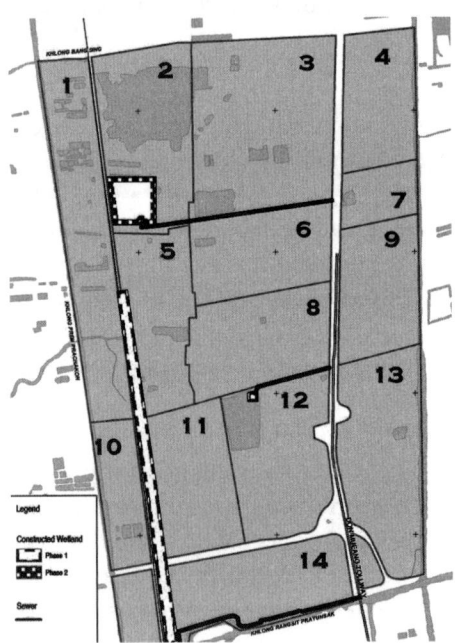

Figure 10-9. Option 2 applied to the pilot area.

of 6,568 m³ per day in the year 2029 (zones 3 and 4). The vertical-flow treatment facility, located on private swampland in the northern part of the subcatchment, would encompass 70,400 m² including treatment plants, infrastructure, and buffer zones; three vertical-flow constructed wetlands; an equalization pond; inlet connections and outlet structures; and construction of the bund (banked containment area).

- *System Southwest:* The southwestern part of the catchment was almost fully developed, and the zones not yet developed were expected to be developed within 10 years. The southwestern part was serviced by three main and two minor wastewater collection systems discharging into Rangsit River through three main and two minor outlets. These collection systems would be utilized as-is. One part of this system covered zones 12 and 14 (Outlets 1, 4, and 5) with an estimated wastewater production of 5,841 m³ per day; another part covered zones 1, 5, 10, and 11 (Outlet 2) with an estimated wastewater production of 12,724 m³ per day, for a total of 18,565 m³ per day in 2029. The proposal was to construct an interceptor (pressure pipe) for zones 12 and 14, while wastewater from zone 11 could be led by gravity through this facility before discharge into Rangsit River. The undeveloped zones 1, 5, and 10 would be connected to the system after 10 years.

Specifications for the vertical-flow treatment facility, located along an 80-m-wide railroad bank in the southwestern part of the subcatchment, included 196,800 m² encompassing treatment plants, infrastructure, and buffer zones. A string of small constructed wetlands, each with an approximately 60 m × 10 m surface area, would be located along the railroad with equalization ponds, inlet connections, outlet structures, and the bund.

We calculated the total investment costs, including the collection system, for the three systems included in Option 2 to be about 230 million baht ($6.9 million USD) with O&M costs at 1.23 baht ($0.037 USD) per m³ treated wastewater.

Option 3: Mini-Treatment. This option involved the installation of mini-treatment plants connected to existing sewer lines at locations where sufficient quantities and qualities of wastewater existed. The wastewater production was similar to the estimates for Option 1. Our proposal was to utilize the existing collection systems, and no new wastewater collection investment would be required for this option (Fig. 10-10).

Because the discharge from the pilot area had a low BOD (about 70 mg/L), we did not consider conventional activated sludge systems to be the most efficient or cost-effective treatment solution. Mini-treatment plants, however, had financial advantages due to their lower requirements for land and no investments in new collection systems needed. Especially for collection systems 2,

Figure 10-10. Option 3: location of mini-treatment plants (black dots) for a key area within the catchment.

4, and 5 in the southern part of the pilot area (covering zones 11, 12, and 14), the use of mini-treatment plants could provide a feasible option because these areas had a high population density, were located close to the outlets (Rangsit River), and lacked open areas suitable for land-based treatment facilities.

The investment specifications for this treatment system included a total of 50 mini-treatment plants, each with a capacity of 600 m^3 per day, to treat 30,053 m^3 per day of wastewater in 2029.

We calculated the total investment costs for Option 3 to be about 300 million baht ($9 million USD), with O&M costs at 0.94 baht ($0.028 USD) per m^3 treated wastewater.

Step 13. Comparisons and Recommendation for the Pilot Area

Our feasibility study analyzed and considered which system, or mix of systems, would be appropriate for the pilot area to service 150,000 persons and cover about 10 km^2. The considerations were based on comparisons and assessment of the technical and financial aspects of the three options. We decided that a mix of Options 2 and 3 would be financially and technically preferable:

- The lake system for the center part—zones 6 through 9—seemed appropriate. This system would require only a small land area in an

area characterized by high land prices; it would utilize an unused lake; it would require only a modest change in the existing collection system; and it would require a very low investment. However, it would incur relatively high O&M costs. Also, this system would require a cooperative agreement between two administrative districts in the catchment.
- The mini-treatment system for the southern part—zones 12 and 14—seemed appropriate. This area lacked land for land-based treatment and would require an interceptor system solution to bring wastewater to a treatment facility upstream. Mini-treatment systems would provide an easy and nonintrusive method to treat wastewater from this developed area
- The vertical-flow wetland system along the railroad for the southwestern part seemed appropriate for the remaining zones. This would entail very low investment and O&M costs and would utilize and beautify a land area that was currently bare.

The approximate investment costs for this preferred combination solution would be in the range of 300 million baht ($9 million USD) and would incur monthly O&M costs of about 700,000 baht ($21,000 USD). These figures turned out to be highly competitive, considering that the investment and O&M activities would end up treating 30,000 m^3 of wastewater per day for about 150,000 PE in 2029. This solution, however, was also rather complicated in terms of planning and implementation, and it depended on a number of nontechnical issues such land availability, cooperation between municipalities, and authority to use government land.

10.1.6 Component 6. Scaling Up: Site Analysis of the Total Project Area

Step 14. Analyze and Estimate Existing Land Use, Population, and Population Equivalent

To assess existing land use, population, and PE, we conducted another satellite imagery analysis. We used the results from the pilot area as a methodological baseline and the project area was divided into four land use categories: residential, commercial, industrial, and open areas, including recreational parks, water-covered areas, and sports facilities.

We subdivided each catchment into these four categories via detailed spatial satellite imagery analysis and we manually counted the number of entities (residential, commercial, and industry units) in each subdivision. This analysis resulted in an estimated PE per land category, catchment, and total. We calculated the spatial areas for each existing category and used them for predictions of future developments.

The total PE for the project area was estimated to be 383,289 in a land area of 95 km^2. Of this, 332,304 were in residential areas (it was estimated that 330,000 people lived in the project area), while the remaining PE derived from the commercial and industrial areas. The PE distribution was very uneven among the catchments, ranging from only 2,990 PE in catchment 2 to 99,728 PE in catchment 4.

Open areas constituted 41% of the total project area, with significant variations among the different catchments (i.e., ranging from more than 70% open area land types in one catchment to only 27% in another catchment).

Step 15. Analyze and Estimate the Existing Infrastructure and Population-Equivalent Served

We surveyed the infrastructure in the pilot area. We extrapolated and combined the results with a number of on-site surveys and interviews with mayors, city clerks, and municipal officers regarding land availability, existing infrastructure, and local recommendations and requirements for future wastewater management. The following general existing infrastructure features in the total project area were highlighted:

- All areas separated grey and black wastewater, including septic tank treatment and seepage of black wastewater.
- The proportion of the PE served by sewer services varied greatly among the different catchments.
- The main discharge outlets were identified; there appeared to be about 40 main outlets in the project area.
- The flow directions were identified but in certain densely populated areas these were hard to locate, even for the municipal officers.

By mapping the service areas, drainage lines, flow directions, and main outlets, we estimated the PE served by sewers in each of the catchments and overall. It appeared that only 55% of the people in the project area were served by wastewater collection services, and the served ratio varied among the catchments, ranging from 0 to 86% (Fig. 10-11).

Step 16. Estimate Future Land Use, Population, and Basic Population-Equivalent

We quantified the future PE through analysis of past land use development, existing land use distribution, and estimated land use development rates for the different catchment areas. If we knew:

- the past development rates for urban developed areas in the project area;
- the existing amount of urbanized areas;

Figure 10-11. Service areas, flow directions, and main outlets in the project area.

- the average wastewater production for urbanized areas based on results from the pilot area study; and
- a qualified estimated development rate for each phase and catchment area, it would be possible to rather precisely predict future PE and thus future wastewater production per catchment area.

Studies of past land use development had shown that the different catchments had developed significant differently during the last 30 years. A previous analysis done by Kasetsart University, using data from 1970 to 2000, showed that development in the province during that period had followed a "1%-5%-3%" development rate ratio (Fig. 10-12 left).

The first development type (the "1%") was characterized by large open areas and little or no infrastructure, especially roads. This type typically had a 1% urban development rate. The second development type (the "5%") was mainly characterized by rapid development of infrastructures, especially roads. This type typically had a 5% urban development rate. The third development type (the "3%") was characterized by well-developed infrastructure and an in-filling type of development. This type typically had a 3% urban development rate.

Based on those development patterns, the existing baseline land use distribution based on satellite imagery, and interviews with key local stake-

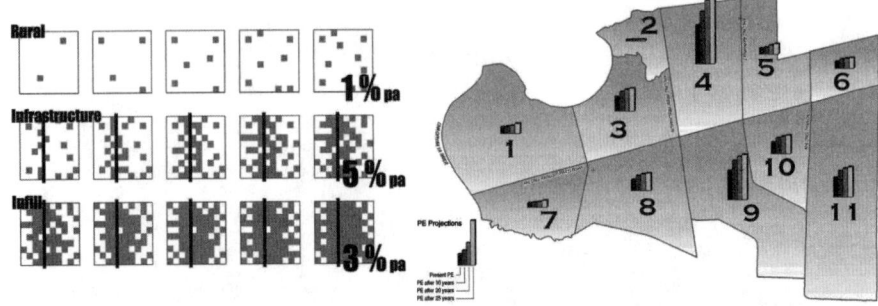

Figure 10-12. The 1%-5%-3% development formula and PE development rates in the different zones of the project area.

pa, per year (annual) growth rate.

holders, we made assumptions for land use development for each of the catchments using the 1%-5%-3% urban development formula. We were then able to calculate the total existing and future PE such that the existing PE of 383,289 was expected to increase in the next 10 years by 31% to 504,018, but then to increase only by 14% and 7% in the following 10- and 5-year periods, respectively, for a final PE of 615,692 in 25 years (Fig. 10-12 right).

Step 17. Adjusted Population-Equivalent Estimate Based on Service Areas

These basic PE figures, however, were too rudimentary and needed to be qualified for existing conditions, existing and predicted service areas, and predicted developments in the 14 different catchments:

- Catchments 1, 2, 3, 5, 6, and 7 were without or almost without wastewater collection infrastructure and the development rates for these catchments were predicted to be low. Based on the expected development ratio in each of these catchments, we estimated future served PE.
- PE for catchment 4, the pilot area, was adjusted according to the detailed prediction made in the pilot study.
- Catchment 8 was well developed but had a below-average connection rate because the catchment was characterized by large housing estates with their own on-site wastewater treatment infrastructure.
- The pilot area study showed that in a densely populated area the connection rate was about 86% of PE. All catchment-served PEs were therefore reduced by a minimum of 14%.

These qualifications resulted in adjusted PE figures per catchment and overall, showing that the existing PE of 383,289, when revised to reflect PE actually served by wastewater collection systems, fell to 211,227. The PE figure was expected to increase in the next 25 years to a final served PE of 510,016 (nonadjusted 615,692).

10.1.7 Component 7. Scaling Up: Apply Options to the Total Project Area

We applied the three options outlined in Step 6 to the full project area and we proposed preliminary design features and outlined estimated costs.

Step 18. Apply Options to the Project Area

Option 1. Catchment Wetland. The catchment wetland option for the project area generally followed the principles applied to in the pilot area (Fig. 10-13). We studied and mapped a proposed layout of the interceptor system and potential sitings of catchment wetlands. Highlights were:

- For wastewater collection, the pressure-pipe option was proposed for all catchments due to financial and technical viability.
- For wastewater treatment, the catchment pond/surface-wetland option was applied to catchments 1, 2, 3, 5, and 7. These catchments were all characterized by low urbanization rates and low land costs, and therefore were more suitable for wastewater management options requiring relatively more land. The catchment vertical-flow wetland option was applied to catchments 4, 6, 8, 9, 10, and 11 for the opposite reasons: higher density, less open space, and higher land prices. We made site visits to all proposed locations for land-based treatment systems, and we found all of the proposed sites to be technically and financially viable. A land cost of about $30 USD/m^2 was found to be the breakeven point for whether a pond/surface-wetland or vertical-flow wetland option should be applied. We preliminarily investigated options for location on government land and found a few, for example, for catchment 8.
- The implementation of catchment wetlands should be phased substantial differently in the various catchments. We proposed that only for catchments 4 and 9 should investments be initiated in the first 10 years. Wastewater management services for catchments 3, 10, and 11 should only be implemented in the following 10 years, while catchments 1, 2, 3, 5, 7, and 8 should be postponed until 20 years hence. (Decentralized, clustered wastewater systems allows for an appropriate phasing of services in a large area, a point worth noting when these systems are compared to conventional centralized systems).
- Total investment costs, including the pressure-based collection system, for the mixed application of pond/surface-flow and vertical subsurface-

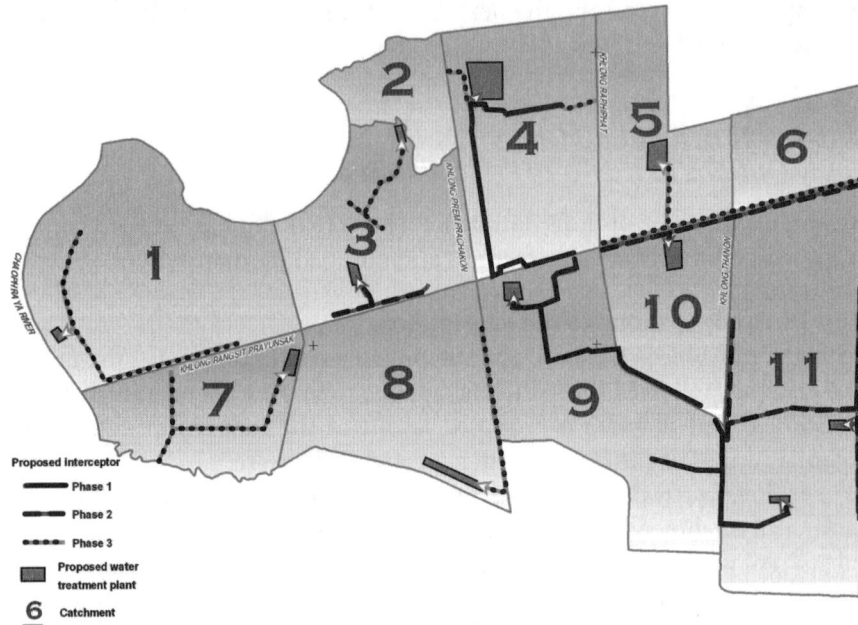

Figure 10-13. Option 1 applied to the project area. Figures correspond to catchment numbers. Catchment 4 compares with the pilot area shown on the left of Fig. 10-8.

flow wetland option for the project area were calculated to be about 1,628 million baht ($48.8 million USD), with an operation and maintenance O&M costs at 0.55 baht ($0.017 USD) per m³ treated wastewater.

Option 2: Cluster Treatment. The cluster treatment option for the project area generally followed the principles applied for the pilot area, but was the hardest to extrapolate because it contained more room for adjustment to local conditions. We did not propose a layout of each subcatchment system, and potential siting for cluster treatment systems were not provided, but we emphasized that the implementation of cluster treatment systems should be phased substantially differently in the various catchments, following the same outline as proposed under Option 1.

Our interviews with key stakeholders provided interesting and useful discussions regarding the possibilities for local subcatchment approaches. Where on the existing drainage system was it possible to intercept the wastewater and gravity-feed it through a wetland? Where were relatively small areas of land available that could benefit from implementation of, for example, vertical-flow constructed wetlands? Was it possible to find vacant government land? The discussions with local key stakeholders opened up new possibilities and options important for finding the best, most suitable local solutions.

We calculated total investment costs, including collection system, for Option 2 to be about 775 million baht ($23.2 million USD), with O&M costs of 0.90 baht ($0.027 USD) per m^3 treated wastewater.

Option 3. Mini-Treatment. The mini-treatment option for the project area followed the principles applied in the pilot area. This option was the easiest to extrapolate because it mainly depended on volumes of wastewater, which were closely linked to PE predictions. We emphasized that implementation should be phased substantially differently in the various catchments, and important issues to be clarified before implementation included actual quantities and qualities of wastewater at the various locations upstream because then the actual number of mini-treatment plants required could be determined. The adjusted PE resulted in a total requirement of 170 mini-treatment plants over the next 25 years. There was a substantial difference of 34 mini-treatment plants [approximately 190 million baht ($5.7 million USD)] between nonadjusted and adjusted PE predictions.

We calculated the total investment costs, including the collection system, for Option 3 to be about 1,015 million baht ($30.4 million USD), with O&M costs at 0.69 baht ($0.021 USD) per m^3 treated wastewater.

10.1.8 Component 8. Results of Comparisons and Analysis

We analyzed the results of our study and compared the three options with each other and with the previous conventional centralized wastewater management study (Table 10-3). The conventional option included main interceptor sewers, covering a length of 60 km, with nine large pumping stations and a single centralized treatment plant with a capacity of 100,000 m^3 per day. We made these comparisons by applying the four proposals to the same project area, the same end-of-project treatment capacity per day, and by using constant prices.

Step 19. Analysis, Comparisons, and Results

Overall, our study found the alternative systems to be appropriate, suitable, financially viable, and applicable in the project area. We concluded that, compared with the previously studied centralized conventional system, the three alternative decentralized options did a much better job of accounting for past management issues (see Step 1 above); they corresponded better to the outlined principles (Step 2); and they fit better with the specific characteristics of the project area (Step 3).

The alternative systems were all cost-effective in terms of investment and O&M. They did not intrude on the existing communities (did not require major new construction works). They took into account local conditions and were linked to previous collection and treatment facility investments in the

Table 10-3. Key Comparison Figures between the Centralized Project and the Three Alternative Options

Key Figures	Centralized	Catchment Wetland	Cluster Wetland	Mini-Treatment
Basics				
Area, km^2	97.00	94.99	94.99	94.99
Population, 1997–2004	150,000	381,339	381,339	381,339
PE served, end Phase 1	163,361	352,599	352,599	352,599
PE served, end Phase 2	269,874	454,894	454,894	454,894
PE served, end Phase 3	310,543	510,016	510,016	510,016
Land cost, USD/m2 1999–2004	60	22.5	22.5	0
System (End Phase 3)				
Sewers, km	60	15	18	0
Treatment capacity, m^3/day	100,000	102,003	102,003	102,003
Land requirements, ha	10.2	108.0	92.2	4.6
Investment Cost (End Phase 3) (million USD)				
Sewer and pumping	47.7	6.8	8.3	0
Sewer major refits	7.4	a	a	0
Land costs (treatment/pump)	6.3	24.4	5.1 c	b
Wastewater treatment plant	31.0	14.2	8.1	28.3
Treatment plant major refits	12.9	a	a	a
VAT	11.9	3.4	1.7	2.1
Total Costs (million USD)	164.9	48.8	23.2	30.4
Project Costs per hectare (million USD/ha)	1.6	0.5	0.2	6.6
O&M Cost (End Phase 3)				
Total O&M Cost (million USD)	47.8	15.4	25.2	19.3
Annual O&M Cost (million USD)	1.6	0.6	1.0	0.8

aIncluded in above.
bUnder roads.
c$18 million USD if no government land is available.
VAT, value-added tax.

area [the septic tanks and local treatment facilities alone were valued at about 250 million baht ($7.5million USD) in the pilot area alone, plus the large investments already made to build well-functioning closed sewer/drainage systems in the densely populated areas].

Even though the alternative systems had major advantages compared to the conventional option, they would be more difficult to plan and conceptualize in detail. However, they avoided the oversimplification and lack of local fit that characterized the proposed conventional wastewater management system.

Our assessments provided the decision makers with a platform for the selection of a wastewater management system but did not provide the final selection because this depended on the province's short- and long-term investment and O&M preferences. How important was it that the system should be the cheapest up-front? How important were the monthly O&M costs when the system was in place? Where was land actually available? How important was the salvage value issue? We emphasized the following key points in our overall assessment and comparisons:

- *Overall investment.* Options 1 and 2, the catchment and cluster wetland options, required only 29% and 14%, respectively, of the conventional project's total investment. The conventional project would cost 5.5 billion baht ($164.9 million USD) compared to 1.6 and 0.7 billion baht for Options 1 and 2, respectively ($48.8 and $23.2 million USD).
- *Land.* One goal of our study was to determine whether land was actually available in the suburban project area. In almost all the catchments land was plentiful and was expected to continue to be so in the next decades. Another goal was to determine whether land-based treatment technologies were technically and financially feasible in the suburban project area. Our cost comparisons showed that they were. If salvage value was included in the feasibility considerations, then land-based options were highly competitive with the conventional wastewater management strategy for the suburban area.

    However, the conventional project would incur only 26% or 34% of the land cost compared to Options 1 and 2, respectively. The conventional system would invest 209 million baht ($6.3 million USD) in 10.2 hectares (102,400 m^2) of land compared to 814 or 599 million baht ($24.4 or $18 million USD) for Options 1 and 2, which needed 108 hectares (1.1 million m^2) and 92 hectares (922,000 m^2), respectively. Option 2 was based on the assumption that no government land was available. If government land was available, Option 2 would only incur 81% percent of the land costs of the conventional project [171 million baht ($5.1 million USD)].
- *Collection system.* Options 1 and 2 required only 12% and 15%, respectively, of the costs for the conventional wastewater collection system. The conventional project would invest 1,834 million baht ($55.1 million USD) in wastewater collection systems compared to only 228 and 278 million baht for Options 1 and 2, respectively ($6.8 million

and $8.3 million USD). The main reason for the cost differences was that the conventional project was predicated on a large, 60-km network of deep-lying gravity concrete sewers that were vulnerable to high groundwater infiltration rates. Option 1 was also based on interceptor sewer lines but used less expensive pressure pipes with lower infiltration rates, and would only collect wastewater from existing or future main outlets. Option 2 was mainly based on use of the existing sewer/drainage system.

- *Treatment systems (not considering land)*. Options 1 and 2 required only 32% and 18%, respectively, of the conventional project costs for construction of treatment facilities. The conventional project would invest 1,465 million baht ($44 million USD) in wastewater treatment systems compared to only 472 and 270 million baht, respectively, for Options 1 and 2 ($14.2 and $8.1 million USD). The main reason for the difference was that the conventional project needed a great deal of expensive electrical or mechanical equipment such as pumping stations and conventional treatment works, compared to the low investment costs for digging, cutting, installing media, and the simple structures in the pond/wetland systems.
- *Short- versus long-term investment.* If short-term investment cost was a main concern, Option 2 was the most cost-effective at about 0.7 to 1.2 billion baht ($23.2 to $36.1 million USD)—about 0.4 to 0.8 billion baht less than Option 1 ($13 to $26 million USD), and 4.5 to 4.8 billion baht less than the conventional project ($129 to $142 million USD).

    Long-term, however, both Options 1 and 2 would be even more cost-effective compared to the conventional project because these options involved purchasing land. Land-based treatment systems are more expensive in the short term but they will in the long term be more financially viable due to the increase in land value (salvage value). For example, land-based systems may after 10 to 20 years be converted to other treatment systems that require less land.
- *Operation and maintenance.* Options 1 and 2 would require only 32% and 52%, respectively, of the conventional project's total investments costs because the conventional project would spend 1,592 million baht ($47.8 million USD) on O&M during the 25-year planned period, compared to 513 and 839 and million baht for Options 1 and 2, respectively ($15.4 and $25.2 million USD). We were not surprised that land-based pond/wetland options would have significantly (30% to 50%) lower O&M costs; we considered this to be very important because the level of O&M costs has a high and very direct influence on system sustainability.

10.2 Reflections on Appropriateness and Sustainability

Two key questions examined by our feasibility study were which technical and financial options are available for large suburban areas, and which methods can be used to undertake preliminary studies of these areas for the large-scale implementation of clustered wastewater management system in developing countries.

The study results provided many valuable experiences for large-scale planning of wastewater management systems in developing countries, and of the planning of decentralized cluster systems in particular. The study also showed that in this type of wastewater management systems planning, conventional wisdom does not suffice. Many of the tools and data usually used, such as relying of secondary data, existing population data, and PE calculation methods, did not provide much help in this context.

10.2.1 Reflections on Technical Appropriateness and Sustainability

Our study showed that contextual, decentralized cluster systems based on gravity flow, pond, and/or constructed wetland techniques can provide an interesting and much cheaper alternative to centralized advanced management systems. It showed that it was possible to assess the feasibility of and plan for cluster systems for large suburban areas close to a metropolitan center to appropriately manage the area's wastewater.

The major focus in the wastewater management system for Pathumthani Province was to start with the natural catchments, thus letting the existing topography and water flows delimit and extend the collection and treatment system. The aim was also to capitalize on what was already in place in terms of existing infrastructures, ditches, pipes, outlets, and local treatment systems, and from there include a mix and match of appropriate wastewater treatment systems customized for each specific catchment.

The study identified three different treatment options. The level of labor intensity, O&M activities, energy supply, mechanization, wastewater composition, land availability, and negotiation of land titles all contribute to the level of complexity of a given treatment system. There is no "ultimate" technical solution. The choice of a decentralized cluster treatment system must be balanced by the abovementioned considerations. Importantly, the issues of re-use and re-entry into the ecosystem should also be considered, but this study did not address them.

An interesting aspect of the use of land-based cluster systems is the question of present costs and future value. For example, after 25 years a conventional system would probably have almost no salvage value. Structures and equipment would have deteriorated and would have to be replaced. The system would go from high investment costs to lower and lower value. For a land-based system, the reverse would be true. In that case, the municipality would invest in land and open spaces (even, as the study showed, for a lower total investment cost than for the conventional systems) and then see the value of this asset go up as the population density and the price of land increase.

Another important aspect is convertability. For a conventional energy-intensive cement system, a conversion to other uses would be almost impossible or at least very expensive. Converting a land-based cluster treatment systems would cause much fewer problems. The land could rather easily be adapted to other uses, such as being sold to private investors (thereby replenishing the municipal coffers) or rehabilitated to function as much-needed open urban space for the benefit of all citizens. The matter of "future-proofing" is relevant in a sector characterized by a general lack of innovation (centralized systems were invented centuries ago) and by a strong need for new, alternative options. For example, a group of experts could suddenly invent a technical improvement to wastewater treatment—perhaps a small, cheap device placed on the pipes to purify the water, making existing capital-intensive systems obsolete.

What seems to be certain is that the wastewater management sector will evolve toward decentralization, ecosystem approaches, and contextual fitting similar to, for example, the energy, water, and solid waste sectors. As a consequence of future requirements for appropriateness and sustainability, the management of wastewater will become more decentralized and will require *more but smaller* wastewater collection, treatment, and re-use systems. This will include the need to minimize the extension of the collection system, to collect wastewater by gravity flow to save energy, to sustainably re-use and re-enter the wastewater into the local ecosystem, to adapt to climate change, and to respond more quickly to urgent needs.

10.2.2 Reflections on Appropriate Planning Methodologies

Our study presented a methodology for analyzing large-scale implementation of catchment and cluster wastewater management options in developing countries. The study design consisted of 8 components divided into 19 analytical steps. These components and steps were relevant, appropriate, and necessary, but they were rather time- and resource-intensive. In many ways they define a "full feasibility" study package, but a broad scope is not always required for appropriate assessments and options. For a more rapid and compact initiative, we found the following lessons learned to be the most valuable and informative.

As always, the establishment of an effective, multidisciplinary team is critical. We found that a team consisting of an experienced, innovative project manager (planner), a constructed-wetland expert, a wastewater collection expert, a GIS expert, and an experienced local engineer (or, perhaps even better, a contractor who can quickly price all the components) constituted a good team. The need for technical conventional wastewater engineers or economists is less important.

Once we had this in place, here are the most important short-cuts and blind alleys we encountered:

Short-Cut 1: Spatial Satellite Analysis. In hindsight, it would have been much more effective to start with satellite images for analyzing a number of key issues, including population figures, land use, spatial distributions of wastewater production, and collection systems. Satellite images, and the accompanying GIS and remote sensing analysis, provided by far the best short-cuts to real-life, factual, and up-to-date data.

Short-Cut 2: Wastewater Flow Analysis. In hindsight, we also should have at the very outset taken wastewater flow data from all of the key outlets as the starting point for analyzing wastewater flows in the catchments, and wastewater quality and quantity. A team on the ground for just a couple of days located all the wastewater outlets from the catchments to the surrounding waterways and conducted 24-hour sampling for each outlet, again providing a short-cut to invaluable real-life, factual, and up-to-date data for the study.

Blind Alley 1: PE Calculations. In hindsight, we should have spent much less time trying to figure out wastewater production by using PE formulas, which contained far too many general, noncontextual assumptions. They mostly resembled a roll of the dice. After unsuccessfully trying to obtain reliable water consumption data from the water companies, we had to wonder whether water consumption should be set at 220 or 250 L/day/PE, and whether the ratio of water to wastewater production should be fixed at 100/80 or 100/75, and so on. It would have been much better to simply go straight to calculating PE figures from the satellite analysis-generated population figures combined with the wastewater flow data we collected, and from our analysis of existing wastewater management infrastructure.

Blind Alley 2: Collecting and Using Secondary Data. We spent too much time, especially in the beginning, on trying to collect secondary data from relevant departments. These data provided by governmental authorities were often difficult to acquire; in many cases they were never received or, in others, were outdated or highly inaccurate. For example, there was a 200% difference between the official population data and the actual number of people found through satellite analysis, house counts, and ground verifications. Looking for secondary data definitely proved to be a blind alley. Clearly, the way to go was to collect and use our own data, as was pointed out in Short-Cuts 1 and 2.

> **Box 10-2. The 8 Study Components and 19 Analytical Steps**
>
> *Component 1. Determine the feasibility options*
> Step 1. Define the typical characteristics
> Step 2. Define the guiding principles
>
> *Component 2. Assess the site characteristics for feasibility options*
> Step 3. Analyze and define the site characteristics
> Step 4. Define the study approach
>
> *Component 3. Define appropriate and sustainable technology options*
> Step 5. Select wastewater management options
> Step 6. Define and outline each option
>
> *Component 4. Scale down before scaling up: site analysis of one catchment*
> Step 7. Determine the present land use
> Step 8. Determine the present population and PE
> Step 9. Determine the present wastewater management infrastructure
> Step 10. Determine the present wastewater production
> Step 11. Predict future wastewater production
>
> *Component 5. Apply alternative systems for the pilot catchment*
> Step 12. Analyze the catchment application for each option
> Step 13. Make comparisons and recommendations for the pilot area
>
> *Component 6. Scaling up: conduct a site analysis of the total project area*
> Step 14. Analyze and estimate the existing land use, population, and PE
> Step 15. Analyze and estimate the existing infrastructure and PE served
> Step 16. Estimate future land use, population, and basic PE
> Step 17. Calculate an adjusted PE based on service areas
>
> *Component 7. Scaling up: apply options to the total project area*
> Step 18. Apply options to the project area
>
> *Component 8. Summarize results of comparisons and analysis*
> Step 19. Present the analysis, comparisons, and results

Short-Cut 3: Simplified Formula for Predicting Future Wastewater Production. It took a long time to figure out a simple method to make predictions of the amount and location of future wastewater production in the different catchments. The method that finally gelled focused on three key questions: (1) how many people there would be in each catchment; (2) how much wastewater they would produce; and (3) how much of this would end up in a sewer. The prediction method we used was thus primarily based on: (1) use of the

1%-5%-3% land development rate formula for phased prediction of future PE; (2) use of the pilot area flow adjusted PE wastewater production formula for estimating future wastewater production in each catchment; and finally, (3) adjusting the wastewater production figures from (2) through a spatial analysis of the expected level of service coverage in each catchment.

With these short-cuts we could finally produce our overall analysis, recommendations, and conclusions. Our focus was on prioritizing the effort. This basically meant that our analysis and recommendations focused, first and foremost, on the financial implications; second, on political opportunities and possibilities; and, third, on technical options. This corresponds pretty well to how real-life decision making is prioritized. However, mainly due to financial constraints and political instability in the country, but also due to the still on-going technical disagreement between proponents of centralized and decentralized systems, no decision has today—5 years later—yet been taken regarding preferred wastewater management system in Pathumthani, and wastewater continues to flow untreated into the canals.

11

The Way Forward, Mainstreaming, and Other Reflections

When did I realize that change for wastewater management in the tropics was not only needed but actually approaching? Personally, when I realized that the tropical dream house I had built had serious flaws! Do not get me wrong—it was a beautifully built house and still is. In the late 1990s I bought a beachfront piece of land, took the previous houses down, designed a new, open-space modern tropical house, and built it together with a local contractor (and his family and his neighbors and their families!). After a year, the Rhom Makham house was finished and fulfilled all my expectations. I applied planar simplicity and transparency to all aspects of the exterior: glazed 6-m-tall walls, thin aluminum frames, lofty white cement pillars, linear white cement roofs, and rectangular structures and attachments. The house is built directly upon and into large natural rocks—rocks that ascend from the beach to about 20 m above sea level. These large, beautifully shaped rocks became an integral part of the exterior and interior of Rhom Makham. And everywhere I sought a balance between the hard-edged materials of marble, steel, and natural rocks and soft materials and spaces for relaxing—comfortable large beds, white furry floor cushions, and soft dining chairs.

To capture and manipulate light and space has always been a quest in modern architecture—shaping houses to receive and display movements of the sun, moon, and clouds; creating lightness and transparency. Pioneer architects like Mies van der Rohe created new ideals for architecture—transparency, material lightness, free-flowing spaces, minimal enclosure, spatial simplicity—that provided the inspirational background for Rhom Makham. Applying these ideals in the tropics was one of my main challenges. How, for example, to let in the hot tropical sunlight and still keep cool? I applied a combination of natural cooling techniques, including the large makham trees located inside the house going through the ceiling to shade the roof; the big windows and doors to let in the sea breeze; the broad marble floors to cool the rooms; and the large, naturally cool rocks (some up to 5 m tall) inside the house as an integral part of all rooms. This experimental combination of natural cooling techniques has proven to create a cool and pleasant living environment in a modern, transparent setting.

Nevertheless, the house is flawed because I followed all the traditional methods when it came to the entire infrastructure. Expensive water must be bought for drinking, toilet flushing, and watering the garden; electricity, also very expensive on this small island, is brought from the grid for lighting, heating hot tap water, air conditioning of the bedrooms, and running the swimming pool and spa pumps; wastewater is treated in a seepage pit that must be relocated about every 5 years; and solid waste is removed and incinerated at the municipal plant. Not exactly impressive. With just a little more awareness and knowledge it would have been easy to make the house much more self-sustaining and eco-friendly. For example, a septic tank with a seepage drain to water the garden, a solar water heater, a couple of solar panels, harvesting water from the 450-m^2 flat roof, composting solid waste, and having a small vegetable garden would have been all that was necessary. There is still time to do these things but it is so much more difficult now compared to during the design phase, when these simple and sensible techniques could have so easily and much more cheaply been incorporated. Irritating. —*Carsten H. Laugesen*

))) 11.1 The Sustainable Ecosystem Approach: Going Mainstream

Most of the case studies in this book were designed, developed, and implemented during the last 5 to 10 years. Still, our reflections upon each of them indicate that some of the issues that were not included—re-use, urban integration, and energy conservation, in particular—have already gained more of a foothold today. Increasingly, they are expected to become more integrated into overall wastewater management systems. The sustainable, ecosystem approach to wastewater management is clearly gaining ground. What only

Figure 11-1. A flawed tropical house?

few years back would have been considered far-fetched, hippie-greenish, something only fanatic environmentalists would consider and advocate, have now become more mainstream and accepted. Intuitively, we see the appropriateness of integration and sustainable local ecosystems, and we wonder why it was not integrated into some of our projects just a few years ago.

This trend is very positive, challenging, and inspiring, and is not only happening within the wastewater management field. The ideas and concepts behind sustainability, decentralized solutions, ecosystem awareness, and the pressures coming from climate change and the need to reduce carbon emissions all point in the same direction. The management of electricity, water, wastewater, and solid waste increasingly becomes interlinked.

Ecological cities are being conceptualized and developed not just at the grass-roots level but also by proactive local and regional governments in both developed and developing countries. Large and ambitious ecosystem and zero-carbon-discharge city projects are being implemented from Denmark to China. Given the growing environmental concern—especially in the light of current discussions on the impacts of climate change—it is likely that houses, settlement clusters, and even cities in 20 to 30 years will fit sustainably into their local ecosystems. Clusters will be developed in existing and new urban areas where energy supply, solid waste, wastewater, sludge, stormwater, water and food supply, and the production of biomass, biofuel, and biodegradable commodities are all being integrated in the design. In addition, treatment units, water retention tanks, and vegetated fields are being linked to a clear social agenda addressing recreational use, formal and informal meeting places, and democratic spaces while still fulfilling a technical function. Technology is just one basic element in the plethora of integrated land use possibilities. In 20 years we might be laughing at the narrow-mindedness and modest level of complexity of the integrated wastewater management systems we are now trying to develop or imagine.

Appropriate management of water and wastewater will continue to be one of the most important keys to sustainability. Climate change will bring water issues to the forefront because the key impacts of climate change will be related to water: more or less water, drought, flooding, water scarcity, growing land competition, migration, poverty—the starting point is water, and the way we manage water and wastewater will become increasingly important. Water and green space have climatic benefits because they reduce temperatures in urban areas through evaporation. In Manchester, UK, for example, surface temperatures have been recorded as 32.1 °C (89.8 °F) in an inner-city square at the same time as 18.4 °C (65.1 °F) in a park. Thus, urban parks have potential as "cooling lungs" providing a broader choice of public spaces for the residents to enjoy. Making the park treat the wastewater, and the water irrigate the trees and re-enter the water cycle through evapotranspiration and percolation, is a natural next step of integrated planning and wastewater management.

In the near future, the linear collection and discharge system will have to be challenged by and changed to cyclic approaches where wastewater is naturally identified and utilized as a resource. Moreover, the time when feasibility studies for wastewater management systems were only based on technical and financial outlines with narrow focus on water is about to end. Sociocultural values, time perspectives, synergy potentials, and general public benefits will increasingly be included as equally important tools in the decision-making process in order to assess the appropriateness and sustainability of a given wastewater management system.

Designing wastewater treatment systems is no longer a task only for civil engineers. Landscape architects, urban planners, ecologists, hotel managers, decision makers, even golfers, school teachers, and sports coaches might join the resources that can and will contribute to the continuous discussion and development of integrated, multifunctional wastewater management systems.

))) 11.2 Three Key Interlinked Conclusions Are Mainstreamed

In this book we have described the state of wastewater management systems in developing countries and provided a framework of 10 guiding principles and 6 elements for appropriate and sustainable wastewater management systems. This framework has been tested in a number of specific cases, varying in scale from a single household unit to a suburban area covering half a million people. The framework is multilayered and each case exhibits different levels of system thinking and success; the knowledge gained from these cases provides valuable insight and experiences for future of wastewater management in developing countries.

Here are our three general, interlinked conclusions for future mainstreaming and action for the development of appropriate and sustainable wastewater management systems in developing countries.

1. The future is not about large-scale, centralized wastewater management but about appropriate, sustainable on-site systems.

The environmental health challenges facing the urban sanitation and wastewater sector in developing countries are two-fold. First, there is the old agenda of providing all urban households with adequate sanitation services. Second, there is the new agenda of managing urban wastewater safely and protecting the quality of vital water resources for present and future populations. The relative importance of each agenda normally depends upon the level of development, although these two agendas coexist in most cities of the developing world, even in some of the most modern cities.

Despite the evident successes of conventional waterborne sewer systems in developed countries, from a sustainability point of view the present concepts of urban, suburban, and rural wastewater management need to be seriously reconsidered in developing countries. Water-based collection systems might, in almost all cases, be inappropriate in the future. In times of scarcity it is detrimental to use up to 70 L of water per person per day just to transport biological matter from our houses to the sea. Even the ivory towers of northern Europe are destined to challenge the extravagant luxury of using potable tap water as a mode of transporting feces.

Clearly, there is an urgent need to improve the sanitation and wastewater management practices in most developing countries. In rural and most suburban coastal areas of developing countries, centralized wastewater collection systems are rarely used; latrines and septic tanks are the most common wastewater disposal systems. These processes can be effective, provided they are designed, installed, maintained, and used properly. A septic tank can remove up to 60% of BOD and suspended solids, and in properly designed septic tanks with soil absorption either through a seepage pit, a drain field, or a constructed wetland, the soil will remove most of the remaining BOD, suspended solids, bacteria, and viruses from the effluent. The biggest problems are lack of de-sludging and improved re-entry systems. Latrines and septic tanks need to be de-sludged periodically or they will result in contamination of the receiving environment, and the sludge must be treated appropriately, such as in a waste stabilization pond or constructed wetland. Seepage pits and drain fields likewise must be constructed and maintained efficiently.

Needed: Successful Large Scale, On-Site Management Systems. All this is well known but rarely implemented in a professional, systematic manner. What is needed today are more successful examples in developing countries of well-functioning, large-scale, on-site management systems. This includes local municipal management of entire on-site systems, from design, construction support, de-sludging, maintenance support, proper disposal of sludge, micro-credit or similar financial support, and so forth. This is the most important area for future support from national, donor, and international financial institutions within the wastewater management sector in developing countries.

2. Keep the focus on on-site systems but improve linkages to cluster and centralized systems.

In some suburban areas on-site system may not be able to stand alone and it may be feasible to develop a local wastewater collection system and use clustered or centralized facilities to treat the community's wastewater. Ponds, constructed wetlands, and sand filters are common, proven, and useful treatment options for medium-sized suburban areas in developing countries. However, as for on-site systems, effluent control practices are normally weak and most of the existing units are today only poorly or not operated and maintained.

Needed: Successful, Large-Scale, Combined On-Site and Cluster Management Systems. What is also needed today are more successful examples in developing countries of well-functioning, large-scale *combined* on-site and cluster management systems. This includes local management, sustainable cluster treatment technology, low-energy consumption, re-use, re-entry, and financial and organizational sustainability. To develop, implement, operate, and maintain such combined systems is the second most important area for future support within the wastewater management sector in developing countries. Such systems could, for example, be developed for a whole province or large municipality. The importance lies in upgrading existing one-off successful demonstration projects to successful, large-scale implementation on a provincial or municipal level. Only in this way will the existing, proven alternative wastewater management systems be able to compete with the conventional centralized systems.

3. The future is not about discharge point, but about land-based wastewater management systems.

For almost all wastewater management systems, the discharge options are very limited. Wastewater streams flow by gravity to point or non-point coastal discharge locations. To avoid excessive electricity costs for pumping, wastewater treatment plants must (and will) be located at the end of these primarily gravity-based wastewater collection systems. Relocation of discharge points is typically not practical, economical, or sustainable. If the environmental sensitivity of the nearby coastal ecosystem is very high, relocation of discharge points might be the only solution, but this would be the exception. The sustainable approach, as highlighted in this book, would be to focus on wastewater management systems that are based on re-use, sustainable re-entry, and low energy consumption.

Needed: Increased Focus on and Experiments with Urban Integrated, Land-Based Wastewater Management Systems. Wastewater management problems are extremely complex and solutions need to be tailored to the specific characteristics encountered in each country, province, and municipality. Proven wastewater management technologies are available but wastewater management systems still seldom include the whole package, from source control, urban integration, and re-use to ecological, sustainable re-entry. What is required are more successful examples of innovative, closed-loop systems that can inspire and challenge the prevailing discharge approach to wastewater management. What is also required are more examples of urban integration of the wastewater management systems: economic, through improved income generation as, for example, in macrophyte-based systems; landscape-wise, through greening and beautification of urban areas; or multifunctional, through combined utilization of the treatment areas for sport, recreation, parking, and so on.

Needed: Treatment of Wastewater for Re-Use. The final result obtained after wastewater treatment is not easily recognized as a valuable product. This explains one of the main reasons why many wastewater treatment systems are poorly maintained and eventually become inactive. If the treatment process itself, in addition to purifying wastewater, could generate valuable products, this would be an important incentive to stimulate optimal operation and maintenance of wastewater management systems. Ecological sanitation aimed at closing the nutrient and water cycles are an interesting example of a re-use system, but many others exist. Large-scale land irrigation for agriculture and forestry are other future-oriented examples. Wastewater re-entry, sludge composting, and biomass production create a "win-win-win" situation while responding to challenges of sanitation and energy supply. Each newly developed wastewater management system, especially if financed internationally, should include and focus on re-use. If this element is not included, the sustainability of the system is doubtful and, at the very least, the system will not add to the required ongoing gathering of experiences with re-use systems. The application of integrated concepts provides a necessary balance between resource utilization, re-use, and environmental protection.

Needed: More Mixing and Matching of Technical Solutions. Technical conventions and standards are usually developed for good reasons and often, when they are promulgated, they embody the technological state of the art. In many cases, however, standards constrain innovation and eventually hinder progress. Innovation and flexibility in technical standards will allow developing countries to expand sustainable access to wastewater management more rapidly and cost-effectively. For example, allowing households, neighborhoods, and communities to choose from a range of technological options based on their preferences and willingness to pay, rather than requiring a uniform standard across an entire city or region, would result in a self-selected technological mix, accelerate progress, and bring improved services to more households in the short term. Decentralizing urban wastewater management planning allows phased implementation of affordable investments within different zones of a city.

Changing technical norms and standards for wastewater services may be challenging, however, because resistance may arise from existing organizations, investors, and technocrats who have a stake in preserving the status quo. It is nevertheless clear that developing countries cannot afford waterborne, sewered sanitation for everyone, and that conventional technologies are not a cost-effective option. What is required is mixing and matching of different alternative wastewater management technologies. In this report, six elements have been defined that should be included in all appropriate and sustainable wastewater management systems in developing countries. This is an area in which international donors and financial institutions have an important role to play by providing acceptance of alternative technologies, local solutions, and alternative appropriate standards.

))) 11.3 Local Context: Going Mainstream

One of the key messages here has been that the local contextual setting is the only framework that can justify a specific technical solution or management setup. There is never just one solution, and only a thorough reading of the local context can indicate, define, or narrow down the scope of appropriate options for that specific site.

Specifically, we would like to highlight four local context issues that should be mainstreamed into future development of appropriate and sustainable wastewater management systems in developing countries.

Context Lesson 1: Local Objectives, Not (Environmental) Standards

Every case study in this book has specific, local wastewater management objectives and treatment goals. For Koh Phi Phi, it was to prevent wastewater from reaching the beach, to reduce odor problems, and to help landowners who could not seep wastewater. For Patong, it was to minimize the amount of raw wastewater reaching the beach, whereas at Siriraj Hospital the absence of adequate public infrastructures supporting the site defined the need for action. To blindly meet standards (uniform, noncontextualized, and politically defined definitions of pollution) was not a primary objective for any of the projects. Likewise, was it not a major priority to improve (either short- or long-term) the water quality in the receiving waters or the conditions for aquatic plants and animals. Only by taking the local defined objectives as the starting point for the design of a wastewater management system can "sense making" and sustainability be taken seriously.

Context Lesson 2: Success and Failure Constantly Interchange, Producing a Rather Difficult Context for Predictions and Lessons Learned

Centralized systems in northern Europe have a long history of sustainable operation and maintenance. Implementing yet another such system in, for example, Hamburg would probably have a pretty good chance of being sustainable. Unfortunately, this is not the case within the wastewater management sector in developing countries. That sphere lacks a history of many successfully operated systems and also lacks consistency regarding the success of individual technologies. To this could even be added the lack of consistency for an individual system over time: what today is a successful, functioning system might tomorrow become a failure, and vice versa—it all depends on some very local and site-specific issues such as political support or interference, or staff mobility.

The cases presented in this book are included partly because they have interesting elements, but mainly because they are tales of personal experience from the field. They are cases in which we or our partners had hands-on experience and followed the process all the way through. In each, we honestly present the successes and failures and discuss the underlying reasons for the outcome. Only by having been there can the context be understood, assessed, and described as real-life stories. They are *not* all success stories from the field, as some of them without doubt will struggle or fail (unfortunately!), while others will succeed and endure. Such case stories are open-ended, where only the future will tell the continued story. Because they are real-life stories and because the contexts have been described as thoroughly as possible, they provide an opportunity for everyone to assess, discuss, and speculate about the appropriateness of the systems and their chances of sustainability. They are put out there in the open for discussion and learning. A revisit in 5 years would probably require some serious rewriting. This is the present state of wastewater management in developing countries.

Context Lesson 3: Tales from the Field as Starting Points for Reflections

That case stories from developing countries are so context-dependent and fluid, however, does not devaluate their use. Rather, it means that their purpose is mainly inspirational, providing starting points for reflections and actions. Case stories make us reflect on basic values, possibilities, and options such as:

- What if policy makers focused on decentralized cluster systems?
- What if wastewater treatment systems were intuitively integrated into the design of holiday resorts, golf courses, public spaces, parks, and motorways?
- What if communities were consequently used as the natural starting points for the implementation of new infrastructures?
- What synergies would arise if the added values of wastewater management systems were identified and could position sanitation as an essential strategic tool to meet broader development goals?
- What if salvage value was included in the feasibility studies of wastewater management systems?

Contextual case stories provide evidence of what is already there in terms of existing systems and technologies, and in terms of new projects and technologies. They also indicate where there is room for improvement and what new knowledge is needed. We hope the framework and cases in this book have been an inspiration, not as a textbook outlining specific options and technologies but as a motivation for "how to think" rather than "'how to do."

An important characteristic of these case studies is that they are attempts to reform or adapt to existing, already implemented infrastructures. This in contrast to, for example, some of the dry systems such as no-flush toilets with composting or incineration units, or flush toilets combined with centrifugal separators, that can be criticized for trying to revolutionize the water and wastewater management systems in a given settlement. It is our experience that the best technical solutions are the ones that reform or adapt to the extant wastewater systems and local knowledge, and that the most cost-effective solution is to utilize these already established investments. In new developments where no previous infrastructures exist, dry systems might be the most feasible option.

Context Lesson 4: Cases of Sense and Simplicity— The Ultimate Way Forward

Sense and simplicity are the suggested two key words to keep in mind when designing a local wastewater management system in a developing country setting. Sense and simplicity have been the guiding principles for the design, implementation, and assessment of the cases presented here.

Sense: The wastewater management system must make sense for local decision makers, financial contributors, taxpayers, citizens, and the community. Making sense is a key factor for local fitness, appropriateness, and sustainability.

Simplicity: Simplicity implies simplifying the management system, the technology, and the details. Simplicity is reducing the level of complexity as much as possible and cutting the number of elements that could eventually lead to the failure of the system. Simplicity does not necessarily make the engineering easier because it normally requires innovative and integrated solutions to create simplicity in design and function. Simplicity is the other key factor for local fitness, appropriateness, and sustainability.

References

Copenhagen Consensus Center (CCC). (2006). "Copenhagen consensus 2006—A United Nations perspective." Outcome paper of meeting on October 27–28, 2006 at UNICEF House, New York, <http://www.copenhagenconsensus.com> (Jan. 29, 2009).

Dreyfus, H. L., and Dreyfus, S. E., with Athanasiou, T. (1986). *Mind over machine: The power of human intuition and expertise in the era of the computer.* New York, The Free Press.

Environment Conservation Department (ECD). (1999). "Survey of wastewater management facilities." Environment Protection Department, Kota Kinabalu, Sabah, Malaysia.

U.S. Environmental Protection Agency (EPA). (2002). *A homeowner's guide to septic systems.* Cincinnati, Ohio, EPA.

Environmental Protection Agency/Queensland Government (EPA/Q). (2001). "Queensland water recycling strategy." Brisbane, Australia, State of Queensland Environmental Protection Agency.

Fujita Research. (1998). "Report on constructed wetlands." <http://www.fujitaresearch.com> (Jan. 29, 2009).

Jastrow, J. (1899). "The mind's eye." *Popular Science Monthly*, 54, 299–312.

Kuhn, T. (1962). *The structure of scientific revolutions.* 1st ed. Chicago, The University of Chicago Press.

Lange, J., and Otterpohl, R. (1997). *Abwasser—Handbuch zu einer zukunftsfähigen Wasserwirtschaft.* Donaueschingen-Pfohren, Germany, Mall-Beton-Verlag.

Laugesen, C. H., and Hansen, J. H. (2003). "Performance improvement report of the wastewater collection system." Phetchaburi Municipality, Thailand, DANIDA/COWI/Lyngby Publishers.

Laugesen, C. H., Paichayoon, S., and Yootana, P. (2004). "Feasibility of wastewater management in nine municipalities in Pathumthani." Bangkok, Thailand, Wastewater Management Authority/ COWI/DANIDA.

Loetscher, T. (1998). *SANEX sanitation expert systems.* Brisbane, Australia, The University of Queensland.

Loetscher, T. (1999). "Appropriate sanitation in developing countries—The development of a computerised decision aid SANEX." Ph.D. dissertation. Brisbane, Australia, Advanced Wastewater Management Centre, The University of Queensland.

Ministry for the Environment, New Zealand (ME/NZ). (2003). *Sustainable wastewater management: A handbook for smaller communities.* Wellington, New Zealand, Ministry for the Environment.

Nelson, M., and Tredwell, R. (2002). "New paradigms: Wastewater gardens, creating urban oases and greenbelts by productive use of the nutrients and water in domestic sewage." *Proc., UNEP Conference on Cities as Sustainable Ecosystems, April 15–16, 2002.* Perth, Australia, Environmental Technology Centre, Murdoch University.

Sawyer, R., ed. (2001). *Closing the loop: Ecological sanitation for food security.* Stockholm, UNDP/SIDA/Water and Sanitation Programme/Thrasher Research Fund/PAHO.

Shaw, R. (1999). *Running water—More technical briefs for health, water and sanitation.* Rugby, UK, ITDG Publishing.

Strauss, M., Heinss, U., and Montangero, A. (2000). "On-site sanitation: When the pits are full—Planning for resource protection in faecal sludge management." *Proc., Int. Conference, Bad Elster, 20–24 Nov. 1998. [Water, Sanitation & Health—Resolving Conflicts between Drinking-*

Water Demands and Pressures from Society's Wastes]. I. Chorus, U. Ringelband, G. Schlag, and O. Schmoll, eds. London, IWA Publishing, WHO Water Series.

United Nations Development Programme (UNDP). (2005). *Health, dignity, and development: What will it take?* Task Force on Water and Sanitation. London, Earthscan.

United Nations Environment Programme (UNEP). (2002a). "A directory of environmentally sound technologies for the integrated management of solid, liquid and hazardous waste for small island developing states (SIDS) in the Pacific region." Suva, Fiji, UNEP/OPUS International/South Pacific Regional Environment Programme/South Pacific Applied Geoscience Commission.

UNEP. (2002b). "Cost benefit analysis case studies in eastern Africa for the GPA strategic action plan on sewerage." In UNEP, *Global programme of action (GPA) for the protection of the marine environment from land-based activities*. Dar el Salaam, Tanzania, UNEP/GPA, University of Dar es Salaam, 2002.

UNEP. (2002c). *Environmentally sound technologies for wastewater and stormwater management—An international source book*. London, IWA Publishing, UNEP–International Environment Technology Centre (IETC), IETC Technical Publications Series 15.

UNEP. (2004). *Improving municipal wastewater management in coastal cities—Training manual*. Nairobi, Kenya, UNEP/UNESCO/IHE.

UNEP. (2006). "Compendium of technologies." UNESCO/IHE/UNEP/GPA Train-Sea-Coast Center, <http://www.training.gpa.unep.org> (Jan. 29, 2009).

van Maanen, J. (1988). *Tales of the field: On writing ethnography*. Chicago, The University of Chicago Press.

World Health Organization (WHO). (1997). *The world health report 1997—Conquering suffering, enriching humanity*. Geneva, WHO.

WHO. (2003). *Looking back, looking ahead: Five decades of challenges and achievements in environmental sanitation and health*. Geneva, WHO.

WHO and United Nations Children's Fund (UNICEF). (2000). *Global water supply and sanitation assessment 2000 report*. Geneva, WHO.

WHO and UNICEF. (2006). *Meeting the MDG drinking water and sanitation target: The urban and rural challenge of the decade*. Geneva, WHO.

Index

activated sludge treatment, 59, 71, 154–55, 157–58
aeration tanks, 3f, 154–55, 160–61
aerobic conditions, 90–91, 98
aero wheels, 154–55, 159–61
  See also rotating biological contactors
aesthetic integration, 18, 60–62, 229
Africa, 4f
agriculture, 65–66
Aikwanich, Pattanapong, 163–64
algae, 180, 189–90
alternative systems, 2–3, 10, 13–16, 229–31
  See also sustainable management
anaerobic conditions, 12, 90–91, 98
animal waste
  biogas digesters, 48f, 49
  fertilizer, 63
anthracite, 54
appropriateness, 19–20, 33–41, 80–87, 229–31
aquaculture, 186, 191–92
aqua privies, 90
Asia, 4f

Ban Pru Teau township project, 100–113
baseline data, 31–32
Bhumibol Adulyadej, King of Thailand, 178–79
biofilters, 54
biogas, 23
biogas digesters, 48f, 49, 61
biosolids, 69
black water, 22, 36
blue-green algae, 180

Canna lilies, 134, 169, 174–76
carbon, 12
case stories, 16–17, 236–37
  alternative collection systems, 44
  Ban Pru Teau township system, 100–113
  coastal ecosystem pollution, 70
  coastal tourism, 39
  combined systems, 58

community management, 74, 76
constructed wetlands, 55, 100–113
ecosystem-focused approach, 35
Koh Phi Phi rehabilitation project, 114–51
malfunctioning centralized systems, 24, 29–30
on-site treatment systems, 47
on-site *vs.* cluster systems, 40
Pathumthani Province planning process, 194–227
Patong river system, 162–77
planning, 31
pond systems, 53
recovery-based closed-loop household system, 89–99
re-entry systems, 68
re-use projects, 64–65
Sakon Nakhon municipal system, 178–93
simplified sewers, 46
Siriraj Hospital system, 152–62
centralized systems, 1–5, 27–30, 41, 52, 233
  capacity, 27
  combined with on-site systems, 57–58
  complexity, 9
  financing, 29
  local input, 27–28
  planning and implementation, 28–29
  politically-based decision-making, 29–30
  recovery-based closed-loop systems, 114–51
  re-entry, 67
  replacement costs, 3
  technical competency, 28
  treatment systems, 52–57
  See also combined systems
central wastewater management agencies, 73–77
centrifugal separators, 42
Channachai, Mr., 180–81
childhood disease, 7
circular disposal solutions, 13–14, 35
circulatory disease, 7, 8f
climate, 20–21, 30–31, 231

closed-loop systems, 12–14, 33, 61
  centralized systems, 123–24, 132–36
  household systems, 89–99
cluster systems, 2, 40–41, 52–57, 230, 233
  constructed wetland systems, 54–56, 100–113
  costs, 71
  overland flow systems, 56–57
  ponds, 52–53
  sand filters, 54
  trickling filters, 53–54
  See also combined systems
coastal ecosystems
  environmental degradation, 176
  pollution, 4, 9, 21, 43, 55, 67, 69–70
  tourism and development, 39, 121
collection systems, 41–45
  alternative off-site systems, 43–45, 46f
  horizontal subsurface-flow constructed wetlands, 105–6
  management at the source, 41–43
  separate wastewater collection systems, 18, 120f, 121–22, 142–47
combined systems, 57–58, 233
  costs, 71
  ponds with subsurface-flow constructed wetlands, 166–77
  ponds with surface-flow constructed wetlands, 18
  septic tanks with drain fields, 50–51
  septic tanks with seepage pits, 49–50, 57, 106
  septic tanks with subsurface irrigation, 96–99
community support, 15–16
  See also local contexts
complexity of existing systems, 26–27
compliance with linkage requirements, 23–24, 36, 147, 165
composting toilets, 42, 48–49
condominium systems, 43–44, 46f
constructed wetlands, 2, 40–41, 54–56
  combined pond and surface-flow systems, 18
  costs, 71
  ecological mechanisms, 176–77
  gravel filters, 134, 135–36, 174, 176
  horizontal subsurface-flow demonstration project, 100–113
  landscaping and vegetation, 132, 169, 171–72, 174–76
  odor reduction systems, 101, 103
  performance data, 58t
  ponds with subsurface-flow systems, 166–77
  ponds with surface-flow systems, 178–93
  public health risks, 110–11
  recovery-based closed-loop centralized system, 123–24, 132–36
  septic tanks with subsurface-flow systems, 51
  shape, 135
  siphons, 141–142
  subsurface-flow systems, 18, 51, 54–55, 132–37
  surface-flow systems, 18, 54, 55f, 56
context. *See* appropriateness; local contexts; sustainable management
cost recovery, 70, 72–73
credit, 79

data, 31–32, 227
death from sanitation-related disease, 7, 8f
decentralized systems, 2–3, 10, 13–16, 70–76
  See also local contexts; sustainable management
deep-sea outfalls, 69–70
denitrification systems, 3f
depth infiltration systems, 54
depth of professional capacity, 82–87
detergent, 42
development funds, 79
diarrhea, 7
disease. *See* public health
disposal technologies, 40
domestic wastewater, 36, 65
  linkage to public networks, 23–24, 36, 147, 165
  off-site systems, 42–45, 46f
  O&M costs, 39
  on-site systems, 4, 5f, 21–22, 36, 41–42
  recovery-based closed-loop system, 89–99
drain fields, 50–51, 68, 71, 90–91
Dreyfuys (Hubert and Stuart) Experiment, 82–86
drinking water, 4, 7–8
dry composting toilets (DCTs), 42
dry systems, 42

dual reticulation, 66–67
duckweed, 53, 61, 192

ecological engineering/ecotechniques, 176–77
economies of scale, 72, 77
ecosystem approach, 10–12, 35
    biogeochemical cycles, 11–12
    constructed wetlands, 176–77
    See also sustainable management
education, 77
educational systems, 9–10
effluent drainage servicing systems, 43–44
effluent taxation, 79
energy requirements, 58–60, 229
    aero wheels, 159, 161
    constructed wetlands, 123
    Patong river system, 171
    pumping stations, 106–7
environmental impact, 9
    See also pollution
environmental standards, 79–80
erosion, 21
Europe, 4*f*
eutrophication, 11–12
evaporation rates, 21
evapotranspiration, 67–68
existing systems, 21–22
    complexity, 26–27, 30
    rehabilitation, 30–31
    simplified sewer retrofits, 45
expertise. *See* technical competency

fecal coliform bacteria, 55
fertilizer, 12–13
financing, 9–10, 15–16
    alternative strategies, 78–79
    capital costs, 71
    centralized systems, 29
    credit, 79
    funding sources, 1
    income/cost recovery, 70, 72–73
    land, 72
    local obligations, 28, 70–73
    operation and maintenance costs, 39, 71, 108–9, 172
first flushes, 21
fishing, 186, 191–92
flooding, 21

flush toilets, 12, 43
    centrifugal separators, 42
    greywater, 51–52
    re-use of treated wastewater, 155–56

Ghana, 24
gravel filters, 134, 135–36, 174, 176
gravity-based systems, 36, 59
grease traps, 27, 36, 92, 143
green algae, 180
greywater, 8, 22, 24–25, 51–52
groundwater contamination, 95

hanging gardens, 95
health. *See* public health
heliconia, 133, 169
horizontal subsurface-flow constructed wetlands, 18, 54, 55*f*, 101–13
    gravel filters, 134, 135–36, 174, 176
    landscaping and vegetation, 133, 169, 171–72, 174–76
    recovery-based closed-loop centralized system, 132–36
    reliability, 134
    shape, 135
    smart technology, 110–13, 173–76
household wastewater. *See* domestic wastewater
human resources, 70

Imhoff tanks, 52
implementation phase, 28–29
income, 70, 72–73
industrial wastewater, 25, 36, 65
infectious disease, 7, 8*f*
institutional systems, 10
integrated wastewater management systems, 18, 35, 57–58, 60–62
International Standards Organization (ISO) 9000 quality certification, 179, 181–83
investment. *See* financing
invisibility, 61
irrigation, 51
    perforated peak-hour pipes, 97
    re-entry systems, 67
    re-use systems, 61, 62–65, 90–92, 155–56
    subsurface irrigation, 17, 67–68, 97
    surface spray irrigation, 69

Jantharo, Witchuda, 117–18, 119f, 127

Kittitarakhun, Phankhum, 118–19, 126–27
Koh Phi Phi rehabilitation project, 114–51, 235
    appropriateness and sustainability, 124–31
    collection system, 120f, 121–22, 142–47
    multifunctional design, 119–21
    O&M, 130–31
    re-use, 123–24
    siphons, 140–42
    smart technologies, 132–51
    solar-powered pumps, 123, 147–51
    subsurface-flow constructed wetlands, 120f, 123–24, 132–36
    urban integration, 138
Komut, Mr., 179, 180f
Kuhn, Thomas, 5–6, 137

lagoons. *See* pond systems
land application systems, 63
land-based re-entry systems, 67–69, 233–35
land-based sludge disposal, 69
land costs, 72
landscaping and vegetation, 17
    subsurface-flow constructed wetlands, 133, 169, 171–72, 174–76
    urban integration, 102–3, 107–8, 110–13
large-scale projects
    alternative planning approaches, 194–227
    on-site treatment systems, 152–62, 233
Latin America, 4f
latrines, 4f, 12, 42–43, 48, 71, 232–33
legislation, 79–80
Limprayoonyong, Niras, 124, 127
linear disposal solutions, 12–13
linkage to public networks, 23–24, 36, 147, 165
loading rates, 27
local contexts, viii, 10–16, 27–28, 33–34, 70–78, 235–38
    biogeochemical cycles, 11–12
    cluster systems, 40–41
    combined systems, 57–58
    commitment and accountability, 75–76, 80
    community support, 15–16
    ecosystem-focused approach, 35
    financing, 78–79
    institutional basis, 76–78
    on-site systems, 232–33
    scale, 40–41
    simplicity, 88, 237–38
    six elements of, 14–16, 33, 34f, 39–41, 81f, 82
    urban integration, 18, 60–62
    See also case stories
low-pressure effluent distribution trenches, 68

macrophytes, 53
Magic, Stella, 159–60
making sense, 237–38
Malaysia, vii
malfunctioning systems, vii, 1–5, 30–31, 116–17, 127–31
management systems, 40–45
    See also sustainable management
manholes, 43
McDonough, William, 95
media. *See* gravel filters
methane gas, 48f, 49, 50, 59, 61
Millennium Development Goals, 7
mini-treatment units, 200–201, 211–12, 219
mosquitoes, 192
multidisciplinarity, 10
multifunctional integration, 61

nitrogen, 12, 43
nod lists
    appropriateness and sustainability, 80–87
    six elements, 13–14, 33, 34f, 39–41, 81f, 82
    smart technology, 17–18, 37
no-flush toilets, 42
Nong Han Lake. *See* Sakon Nakhon municipal system
non-water-based systems, 42
North America, 4f

Oceania, 4f
odor reduction systems, 101, 103, 107–8, 192
off-site systems, 42–45, 46f
oil traps, 27, 36, 92, 143
on-site systems, 232–33
on-site systems for households, 4, 5f, 36, 41–42

dry systems, 42
failures, 57
inspections, 47
latrines, 4f, 12, 42, 43
O&M, 39, 57, 71
recovery-based closed-loop system, 89–99
re-entry, 67
septic tanks, 4f, 21–23, 43
soak-away systems, 105
treatment, 45–52
zero discharge approach, 89–99
See also combined systems
on-site systems for large facilities, 152–62, 233
operation and maintenance
   activated sludge treatment, 158
   costs, 39, 71–73
   energy requirements, 58–60
   inspections, 47
   local obligations, 28, 70–76
   on-site systems, 39, 47, 57, 71, 162
   ponds and surface-flow constructed wetlands, 187–88
   pumping systems, 146
   recovery-based closed-loop systems, 123–24, 130–31
   siphons, 140
   sludge, 162
   solar-powered pumps, 148–51
   subsurface-flow constructed wetlands, 106, 108–9, 172
organic materials, 12
overhead sprinklers, 67
overland flow systems, 56–57
oxygen transfer, 12

Pak Bang River and Pak Lak Canal. *See* Patong river system
paradigm shifts, 5–6, 136
parasitic disease, 7, 8f
Pathumthani Province planning process, 194–227
   appropriateness and sustainability, 223–25
   area assessments, 196–98
   comparisons with conventional systems, 219–23
   complete site analysis, 213–17
   complete systems analysis, 217–23
   feasibility assessments, 195–96
   pilot site analysis, 201–7
   pilot systems analysis, 207–13
   review of planning process, 224–27
   selecting technology options, 198–201
Patong river system, 162–77, 235–36
   appropriateness and sustainability, 169–72
   energy use, 171
   inlet collection structure, 167–68, 171, 173–74
   O&M, 172
   original system, 164–65
   outlet system, 169
   ponds with subsurface-flow constructed wetlands, 166–77
   urban integration, 171–72
PE (Population-Equivalent) methodology, 205–7, 225–26
perforated peak-hour pipes, 97
phosphorus, 12, 42, 43
Phra That Choeng Chum temple, 178–79
pipe biofilm reactor. *See* aero wheels
pit latrines. *See* latrines
Planck, Max, 6
planning phase, 1–2, 28–32
political dynamics, 10
   commitment and accountability, 75–76, 80
   instability, 80
   preferences for centralized systems, 1
pollution, 8–9
   coastal ecosystems, 4, 9, 39, 43, 176
   effluent taxation, 79
   eutrophication, 11–12
   industrial wastewater, 65
   organic materials, 12
   storm drains, 24–25
   toxic materials, 42, 65
pond systems, 3f, 52–53
   algae, 180, 189–90
   combined with subsurface-flow constructed wetlands, 166–77
   combined with surface-flow constructed wetlands, 18, 178–93
   costs, 71
   multifunctional integration, 61
   plant material, 53
pour-flush latrines, 48, 71
poverty, 7–8

professional capacity. *See* technical competency
promotion and education, 77
public health, 7–9, 10
    constructed wetlands, 110–11
    re-use systems, 63–65
    water-borne disease, 7, 8*f*, 21
pumping, 36
    horizontal subsurface-flow constructed wetlands, 103, 106–7
    O&M, 146
    solar-powered pumps, 18, 59–60
    vs. siphons, 141–42

rainfall, 20–21, 24–26
rapid infiltration systems, 69
recovery-based closed-loop systems
    centralized systems, 114–51
    household systems, 89–99
recovery of energy, 59
re-entry, 67–70
    cluster systems, 52
    land-based systems, 67–69
    overland flow systems, 56–57
    water-based systems, 69–70
regulation, 77–80
rehabilitation of existing systems, 30–31
research centers, 184
re-use systems, 15–16, 35, 48–49, 61–67, 229–35
    business and industrial uses, 65
    composting toilets, 48
    domestic uses, 66–67
    environmental uses, 67
    fertilizer, 12–13, 63, 69, 91
    flush toilets, 155–56
    horizontal subsurface-flow constructed wetlands, 107
    irrigation, 51, 61–66, 90–92, 155–56
    pollution, 65
    public health risks, 63–65
    recovery-based closed-loop centralized system, 123–24
    recreational uses, 67
    sand filters, 54
    sludge, 23
    urine-diversion toilets, 48–49
river treatment, 170
    See also Patong river system

robustness, 37
rotating biological contactors (RBCs), 18, 54, 91–92, 97–98, 154–55, 159
rotating perforated tubes (RPT). *See* aero wheels
rules, 82–83

Sakon Nakhon municipal system, 178–93
    appropriateness and sustainability, 185–88
    aquaculture, 186
    collection system, 185
    energy use, 186
    ISO 9000 certification, 179, 181–83
    O&M, 187–88
    ponds and surface-flow constructed wetlands, 180–81, 185–86
    research center, 184
    smart technology, 188–93
    urban integration, 186–87
    visitors' center, 182–84
sand filters, 51, 54, 58*t*
sanitation credit funds, 79
sanitation-related disease, 7, 8*f*
scale, 40–41
seepage systems, 21–22, 57, 67
    combined with septic tanks, 49–50, 106
    costs, 71
    gravel filters, 134, 135–36
    groundwater contamination, 95
    subsurface systems, 68
    surface systems, 69
sense, 237–38
separate wastewater collection systems, 18, 142–47
septic tanks, 21–22
    cleaning, 47, 50, 98
    combined with alternative offsite collection systems, 43
    combined with drain fields, 50–51
    combined with seepage pits, 49–50, 57, 106
    combined with subsurface-flow constructed wetlands, 51
    combined with subsurface irrigation, 17
    costs, 71
    nitrogen and phosphorus removal, 43
    performance data, 58*t*
    recovery-based closed-loop systems, 89–99
    re-entry, 67

regions using, 4f
sludge, 22–23, 232–33
settled sewage systems, 43–45
sewers, 4f
    costs, 71
    domestic collection, 42–43
    manholes, 43
    re-entry, 67
    simplified systems, 43–45, 46, 71
    slope, 45
short systems, 36
simplicity, 37, 87–88, 237–38
simplified sewers, 43–45, 46, 71
siphons, 18, 60, 98, 141–142
Siriraj Hospital system, 152–62, 236
    activated sludge treatment, 154–55, 157–60
    aero wheels, 154–55, 159–61
    appropriateness and sustainability, 156–59
    collection, 154, 157
    energy use, 159, 161
    re-use, 155–56
    sludge removal, 162
    smart technologies, 159–62
    urban integration, 159
sludge fertilizer, 91, 98
sludge treatment systems, 2, 22–23, 232–33
    costs, 71
    energy requirements, 59
    land-based systems, 69
    mini-treatment units, 200–201, 211–12, 219
    ponds, 52
small borehole systems, 43–44
smart technology, 17–18, 37
    activated sludge treatment, 159–62
    combined pond and constructed wetlands, 188–93
    horizontal subsurface-flow constructed wetlands, 110–13, 173–76
    recovery-based closed-loop centralized systems, 132–51
    separate wastewater collection systems, 142–47
    septic tanks with subsurface irrigation, 96–99
    siphons, 140–42
    solar-powered pumps, 147–51

soil erosion, 21
solar-powered systems
    battery costs, 149–51
    integrated photovoltaic cells, 59, 151
    pumps, 18, 59–60, 147–51
Somchat, Mr., 180f, 182–84
Sommai, Mr., 118–19, 127
spatial satellite analysis, 225
stalled treatment units, 3f
stormwater, 24–26, 36
The Structure of Scientific Revolutions (Kuhn), 5–6
submerged contact biodisc aerator (SCBA). *See* aero wheels
subsurface-flow constructed wetlands, 5, 54–55
    combined with ponds, 166–77
    combined with septic tanks, 51
    gravel filters, 134, 135–36, 174, 176
    landscaping and vegetation, 133, 169, 171–72, 174–76
    recovery-based closed-loop centralized system, 132–37
    shape, 135
subsurface irrigation, 17, 67–68, 97
subsurface seepage systems, 68
surface dripline irrigation, 69
surface-flow constructed wetlands, 18, 54, 55f, 56, 178–93
surface seepage systems, 69
surface spray irrigation, 69
sustainable management, 9–32, 229–31
    appropriateness, 19–20, 33–41, 80–87
    biogeochemical cycles, 11–12
    closed-loop systems, 12–15, 33
    decentralized systems, 2–3, 10, 13–16, 70–76
    economies of scale, 72
    ecosystem approach, 10, 11–13
    energy requirements, 58–60
    financial feasibility and autonomy, 39, 70–73, 78–79
    levels of professional capacity, 81–87
    on-site systems, 232–33
    re-entry into the ecosystem, 67–70
    regulation, 79–80
    re-use of human waste, 12–13, 63
    re-use of treated wastewater, 15–16, 62–67

sustainable management, (continued)
  scale, 40–41
  simplicity, 37, 87–88
  smart technology, 17–18, 37
  temperature and climate, 20–21
  urban integration, 60–62
  See also local contexts
Suwarnarat, Ksemsan, 89–99, 156
symbolic integration, 61
system thinking, 10

technical competency, 1, 28, 72, 77, 82–87
temperature, 20–21
Thailand, 19
  Ban Pru Teau township demo project, 100–113
  centralized systems, 2, 3f, 29–30
  climate, 21
  Koh Phi Phi rehabilitation project, 114–51
  malfunctioning systems, vii, 116–17, 127–31
  Pathumthani Province planning process, 194–227
  Patong treatment system, 162–77
  planning process, 31
  recovery-based closed-loop household system, 89–99
  Sakon Nakhon municipal system, 178–93
  Siriraj Hospital system, 152–62
  soil erosion, 21
  suburban drain systems, 26
  tsunami of 2004, 100, 114–15, 118
Todd, John, 95
toilet paper, 97
toilets
  aqua privies, 90
  biogas digesters, 48f, 49
  composting, 42, 48–49
  flush toilets, 12, 42, 43
  greywater, 51–52
  no-flush toilets, 42
  re-use of treated wastewater, 155–56
  urine-diversion, 48–49
topographic integration, 62
tourism, 39, 121, 163–65
treatment systems, 14–16, 45–58
  cluster systems, 52–57
  combined systems, 57–58
  costs, 71
  large-scale systems, 152–62
  urban integration, 60–62
  See also centralized systems; on-site systems
trench systems, 68
trickling filters, 53–54
tropical fitness, 30–31
tsunami of 2004, 100, 114–15, 118

United Nations Millennium Development Goals, 7
urban contexts, 26–27, 60–62
  aesthetic integration into the environment, 18, 60–62, 229
  alternative off-site collection systems, 43–45, 46f
  centralized wastewater management, 76–78
  combined approach, 57–58
  evolution, 24–26
  landscaping, 102–3, 107–8, 110–13
  linkage from households to public networks, 23–24, 36, 147, 165
  planning and feasibility studies, 194–227
  zero-discharge systems, 94–95
  See also centralized systems
urine-diversion toilets, 48–49

vegetation. See landscaping and vegetation
ventilated improved pit privies (VIPs), 48, 71
vertical subsurface-flow constructed wetlands, 18, 54, 55f
  efficiency, 134
  recovery-based closed-loop centralized system, 132–36
  shape, 135
  siphons, 140–142
visitors centers, 182–84

waste stabilization ponds. See pond systems
wastewater. See black water; collection systems; domestic wastewater; greywater; industrial wastewater; treatment systems
wastewater flow analysis, 225

wastewater production estimates, 205–7
water-based re-entry systems, 68, 69–70
water-borne disease, 7, 8f, 21
water hyacinth, 53, 192
water supply, 4, 7, 8f
water utilization, 14–16
wetlands, 20f, 67
　See also constructed wetlands

width of professional capacity, 81–82, 84f
wind power, 151
Wines, James, 95
The World Bank, 75

Yeahg, Ken, 95

zero discharge systems, 89–99

About the Authors

Carsten Hollænder Laugesen is a senior international development specialist. Presently he is Development Counselor at the Royal Danish Embassy in Pretoria in charge of all Danish development assistance to southern Africa. For one of Europe's leading consultancy firms, COWI A/S, he has been chief technical advisor and project manager on numerous projects. Involved in the design, appraisal, implementation, and evaluation of more than 80 national and international development projects during the last 20 years, Mr. Laugesen has acquired comprehensive knowledge of project management and methods for innovation, implementation, problem analysis, surveying, and evaluation. His professional experiences extend to 16 different countries, and for the last decade he has been permanently residing in Southeast Asia and South Africa. Through practical experiences on the ground with physical, social, financial, and organizational issues, Mr. Laugesen has developed an intuitive understanding of which parameters make development assistance work—and fail. Mr. Laugesen is the initiator and the driving force behind this book. He is a diligent team leader, an innovator, and a broad-minded intellectual who envisions projects and motivates everyone involved to do their utmost to reach a common goal. Mr. Laugesen moves quickly, is an excellent interpreter of a given contextual setting, and is a professional who can assess and act with a unique combination of knowledge, experience, and intuition. *An empathetic and humorous professional.*

Ole Fryd is trained as an architect and urban planner and works as a researcher in integrated water management and urban development at The Danish Centre for Forest, Landscape and Planning (Forest & Landscape Denmark) within the University of Copenhagen. During past years he has lived and worked as a researcher, planner, and consultant in Denmark, Greenland, and Thailand, and has traveled extensively to an additional 50 countries. He joined the team to develop new ways of integrating utility networks into physical urban design and to reflect upon the level of contextual fitting seen in wastewater management systems in Southeast Asia. Mr. Fryd steadily advocates for an interdisciplinary and context-based approach to urban environmental challenges, and tries to promote a better way forward in public administration and in academic and commercial circles though seminars, workshops, papers, books, and practical projects on the ground. *A newcomer with a mission.*

Thammarat Koottatep is an Assistant Professor in Environmental Engineering at the Asian Institute of Technology in Klong Luang, Pathum Thani, Thailand. He is a leading expert in decentralized wastewater treatment systems and the application of eco-engineering technologies in Southeast Asia. His research activities include the application of constructed wetlands for septage treatment and the development of sustainable on-site wastewater treatment systems for small-scale communities. Dr. Thammarat has a broad international perspective, a vast professional network, and unprecedented insight into professional and nonprofessional relations defining successes in the Southeast Asia region. He is a unique bridge-builder linking Western and Eastern traditions and Northern and Southern agendas in the implementation of environmental management projects. *A bright and soft-spoken gentleman.*

Hans Brix is a Professor of Plant Ecophysiology at the Department of Biological Sciences, Aarhus University, Aarhus, Denmark. His specializations are freshwater ecology, wetland ecology and management, and the implementation of wetland systems for water pollution control. For two decades Dr. Brix has been one of the world's leading pioneers in the use of constructed wetlands for treatment of wastewater. His research involves studies on the release of oxygen from plant roots and physiological characteristics of aquatic plants in relation to growth conditions (nutrient uptake). He utilizes this expertise to optimize treatment efficiency in constructed wetlands through the choice of plants, growth media, water depth, and so forth. His research has been published in more than 150 scientific papers in journals, books, conference proceedings, and reports. *A warm-hearted idealist.*